Experiments in Electricity 3E

For Use with the Lab-Volt® EMS Training System

Stephen L. Herman

Australia Canada Mexico Singapore Spain United Kingdom United States

Experiments in Electricity for use with Lab-Volt® EMS Equipment, Third Edition

Stephen L. Herman

Vice President, Technology and Trades SBU:
Alar Elken

Editorial Director:
Sandy Clark

Acquisitions Editor:
Dave Garza

Development Editor:
Jennifer A. Thompson

Marketing Director:
Cyndi Eichelman

Channel Manager:
Fair Huntoon

Marketing Coordinator:
Brian McGrath

Production Director:
Mary Ellen Black

Production Editor:
Toni Hansen

Art/Design Specialist:
Mary Beth Vought

Senior Editorial Assistant:
Dawn Daugherty

Cover Design and Illustration:
John Kenific

Printed in the United States of America

6 7 8 9 XX 09 08 07

For more information contact
Delmar Learning
Executive Woods
5 Maxwell Drive, PO Box 8007
Clifton Park, NY 12065-8007
Or find us on the World Wide Web at
www.delmarlearning.com

Library of Congress Cataloging-in-Publication Data:
Card Number: 2003055114

ISBN-13: 978-1-4018-2563-8
ISBN-10: 1-4018-2563-X

NOTICE TO THE READER

Contents

Preface

The experiments in this laboratory manual are designed to be used with the Lab Volt® EMS (Electro-Mechanical System) training system. These experiments were tested by students for several years in a laboratory environment using the EMS equipment. They are intended to give the student hands-on knowledge of different types of measuring instruments and electrical equipment. Although most of the experiments employ the equipment found in the EMS system, there are a few experiments that require additional equipment, such as an ohmmeter or an extra ammeter. The student is led through a step-by-step procedure for computing values of different types of electric circuits, and then these values are proven by connecting the circuit and using meters to measure the computed values. Some of the basic types of electrical circuits and machines covered in this manual are:

1. Series, parallel, and combination circuits.
2. Resistive, inductive, and capacitive AC circuits.
3. RLC series and parallel circuits.
4. Single-phase transformers.
5. Three-phase circuits.
6. Power factor correction for both single-phase and three-phase circuits.
7. Three-phase transformer connections.
8. Single-phase loads for three-phase transformers.
9. Direct current generators and motors.
10. Three-phase alternators.
11. Three-phase motors.
12. Single-phase motors.

In the circuit connections presented in this manual, ammeter and voltmeter range values are given for the highest expected readings during the experiment. **These are not the voltages that are to be applied to the circuits.** Beginning students often confuse the range of a meter with the amount of voltage that is to be applied to the circuit. The meter range indicates the maximum voltage, current, power, etc., that the meter is capable of safely measuring. During an experiment, it may be necessary to change the meter's range setting. As a safety precaution, the student should turn off the power to the circuit before changing the range setting of a meter.

Since the first edition of this text, Lab Volt® has made changes to some of the modules used in the EMS system. The power supply, for example, now has a 24 VAC outlet to power certain meters that require external power. Wattmeters now require this external power to operate. These changes will be noted in the text. Also, another piece of equipment called the "prime mover/dynamometer" is now available that was not available when this text was first written. This module can be used as either a prime mover to power generators, or as a dynamometer to provide a load for motors. One advantage of this module over the older dynamometer is that it provides a digital display that indicates torque and speed.

The basic format of this manual follows that of ***Delmar's Standard Textbook of Electricity 3e*** and is intended to be used with that textbook. The lab manual has been written as a stand-alone book, however, and is not dependent on the information contained in that textbook. This manual can be used with any textbook on the subject of industrial electricity.

NEW FOR THE THIRD EDITION

Since this book was first published, Lab Volt® has produced new equipment that was not covered in the previous editions. Probably the most important to this manual is the EMS model 8960-10 prime mover/dynamometer, as shown in Figure P-1. The prime mover/dynamometer module can be used to replace other motors that are employed as prime movers to operate generators, or to replace the EMS 8911 electrodynamometer module. When this module can replace the one specified in the text, a notation will be made to this effect.

Using the EMS 8960-10 as a Dynamometer

When using the EMS 8960-10 as a dynamometer, it is necessary only to connect the DC machine to the motor being tested with a timing belt, connect the 24 VAC low-power input to the 24 VAC supply on the power supply, and set the MODE switch in the DYN. position. The MODE switch, located inside the DYNAMOMETER LOAD CONTROL box, is generally set in the manual (MAN.) position. This permits the manual control knob to adjust the amount of torque produced by the dynamometer. If the display switch is set in the TORQUE position, the display will indicate the torque in pound feet/inch (inch-pounds of torque).

Using the EMS 8960-10 as a Prime Mover

If the EMS 8960-10 is to be used as a prime mover, a timing belt is again used to connect the DC machine of the prime mover to the generator or alternator. The PRIME MOVER INPUT jacks are connected to terminals 7 and N of the power supply. Terminal 1 of the prime mover is connected to terminal 7 of the power supply and terminal 2 is connected to terminal N of the power supply. The 24 VAC power input must again be connected to the 24 VAC supply of the power supply. The MODE switch should be set in the PRIME MOVER position and the DISPLAY switch should be set in the SPEED position. The display will now indicate the speed of the prime mover. The amount of DC voltage applied to the prime mover by the power supply determines the speed of the prime mover.

The module also contains TORQUE OUTPUT, SPEED OUTPUT, and SHAFT ENCODER OUTPUTS. These outputs produce a voltage that is proportional to the torque or speed of the DC machine.

FIGURE P-1 EMS 8960-10 Prime mover/Dynamometer (Courtesy of Lab-Volt® Systems, Inc.)

About the Author

Stephen L. Herman has been both a teacher of industrial electricity and an industrial electrician for many years. His formal training was obtained at Catawba Valley Technical College in Hickory, North Carolina. Mr. Herman has worked as a maintenance electrician for Superior Cable Corp. and as a class "A" electrician for National Liberty Pipe and Tube Co. During those years of experience, Mr. Herman learned to combine his theoretical knowledge of electricity with practical application. The books he has authored reflect his strong belief that a working electrician must have a practical knowledge of both theory and experience to be successful.

Mr. Herman was the Electrical Installation and Maintenance instructor at Randolph Technical College in Asheboro, North Carolina for nine years. After a return to industry, he became the lead instructor of the Electrical Technology Curriculum at Lee College in Baytown, Texas. He retired from Lee College after 20 years of service and, at present, resides in Pittsburg, Texas, with his wife, Debbie. He continues to stay active in the industry, write, and update his books.

Acknowledgments

The author and publisher wish to thank the following individuals for their technical review of the content:

Michael Turner
Central Carolina CC
Sumter, SC

Max Holland
Chattanooga State Tech CC
Chattanooga, TN

Safety

Objectives

After studying this unit the student should be able to:

- Discuss basic safety rules.
- Describe the effects of electric current on the body.

The laboratory experiments contained in this text are designed to be used with the Lab-Volt EMS training system. The Lab-Volt EMS system is a hands-on device designed to train students in the industrial electrical maintenance field. For this reason, much of the equipment is designed to be operated at rated line voltages of 120 or 208 V. This equipment has been engineered to eliminate the hazard of electrical shock and mechanical injury as much as possible, but the student should be aware that it is impossible to eliminate all shock or mechanical hazards from this type of equipment. The primary responsibility for safety must, therefore, be assumed by the person using the equipment.

GENERAL SAFETY RULES

Think

Of all the rules concerning safety, this one is probably the most important. No amount of safe-guarding or "idiot proofing" a piece of equipment can protect a person as well as the person taking time to think before acting. Many electricians have been killed by supposedly "dead" circuits. Do not depend on circuit breakers, fuses, or someone else to open a circuit. Test it for yourself before you touch it. If you are working on high voltage equipment, use insulated gloves and meter probes designed to be used on the voltage being tested. Your life is your own, so think before you touch something that can take it away from you.

Avoid Horseplay

Jokes and horseplay have a time and place, but the time or place is not when someone is working on a live electrical circuit or a piece of moving machinery. Do not be the cause of someone being injured or killed, and do not let someone else be the cause of your being injured or killed.

Do Not Work Alone

This is especially true when working in a hazardous location or on a live circuit. Have someone with you to turn off the power or give artificial respiration. One of the effects of severe electrical shock is that it causes breathing difficulty.

Work with One Hand When Possible

The worst condition for electrical shock is when the circuit exists between two hands. This causes the current path to pass directly through the heart. A person may survive a severe shock between the hand and one foot that would otherwise cause death if the current path was from one hand to the other hand.

Learn First Aid

Anyone working in the electrical field should make an effort to learn first aid. This is especially true for those electricians who work in an industrial environment and are exposed to high voltages and currents. A knowledge of first aid may save your life or someone else's life.

An Analogy of an Electrocution

The following is a description of the effects of different values of electric current on the body.

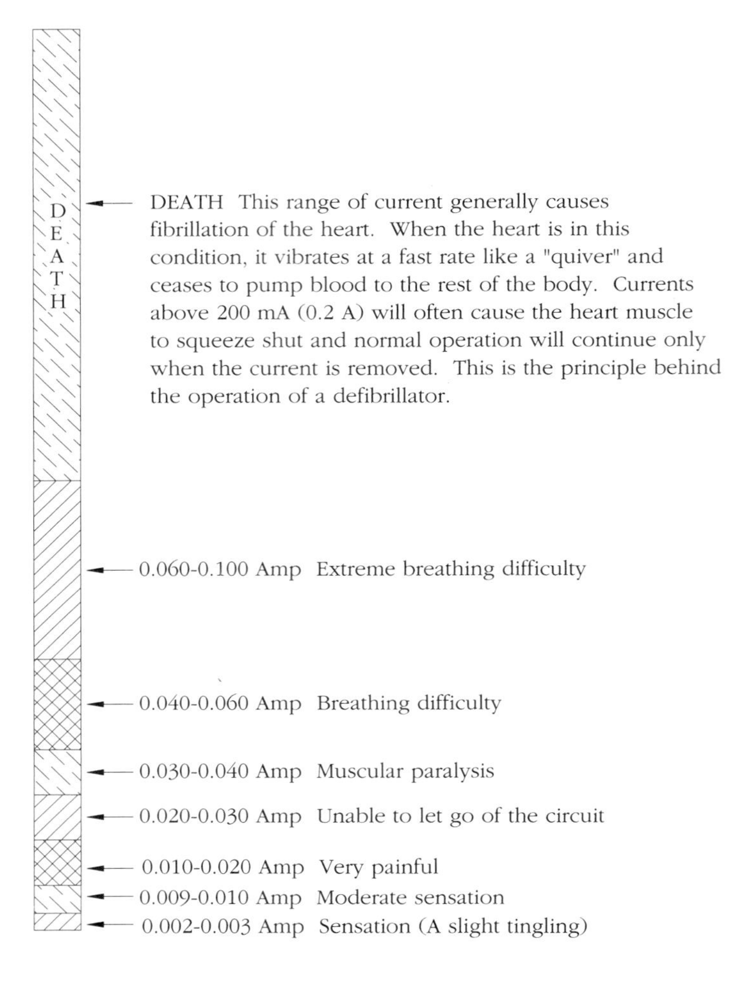

Review Questions

1. What is the most important rule of safety?
2. Why should a person work with only one hand when possible?
3. What range of electric current generally causes death?
4. What is fibrillation of the heart?
5. On what principle of operation does a defibrillator work?

Exercise 1

The Power Supply

Objectives

After completing this lab you should be able to:

- List rated voltages for different terminals on the power supply.
- Measure voltages at different terminals on the power supply.
- Discuss the operation of voltmeters and ammeters.
- Connect voltmeters and ammeters.

Materials and Equipment

Power supply module	EMS 8821
Resistance load module	EMS 8311
DC metering module	EMS 8412
AC voltmeter module	EMS 8426
AC ammeter module	EMS 8425

Discussion

The 8821 power supply module is the heart of the lab volt EMS (electromechanical system) system. It is used to supply all power to operate the various modules used to perform the experiments in this manual. It is necessary to become familiar with the different voltages and current ratings of the power supply.

Newer models of the 8821 power supply contain a 24 VAC output that is used to power other pieces of equipment such as wattmeters and the prime mover/dynamometer module. Students will be instructed to make connection to this power output when necessary throughout the text. An 8821 power supply model is shown in Figure 1-1.

Caution *should be exercised when connecting and disconnecting any circuit, because the EMS system uses full line voltages of 208/120 V. These voltages are high enough and the current ratings are sufficient to cause a very dangerous electrical shock. Although the system has been designed to reduce the electrical shock hazard, no amount of engineering and design can substitute for good safety habits.*

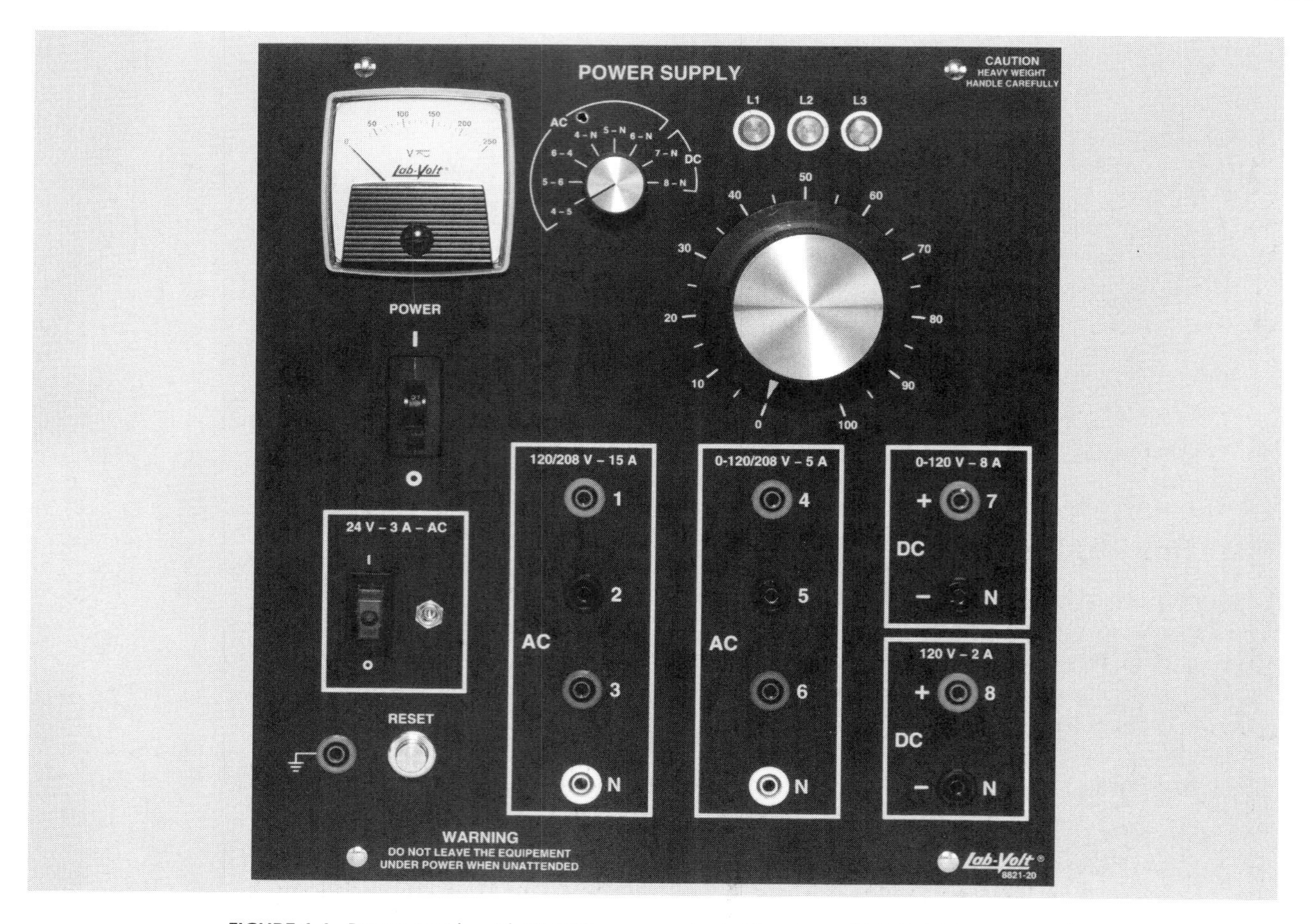

FIGURE 1-1 Power supply with 24 VAC power output. (Courtesy of Lab-Volt® Systems, Inc.)

Name ______________________________ Date ______________

Procedure

1. Examine the front of the power supply module and locate the main power switch. This switch is located below the meter. The main power switch is used to disconnect all power from the module.

2. Locate the terminals marked 1, 2, 3, and N. Notice that these terminals are surrounded by a separate box. About midway down the box are the letters AC. This indicates that the voltage supplied by these terminals is alternating current. At the top of the box the voltage and current rating for these terminals are shown (120/208 V–15 A). These terminals will provide 120 V AC between any of the 1, 2, or 3 terminals and N. 208 V is provided between any of the 1, 2, or 3 terminals. These terminals have a maximum current rating of 15 A. These terminals also provide a "fixed" voltage. This means that the voltage cannot be adjusted or changed.

3. Locate the terminals labeled 4, 5, 6, and N. These terminals are inside a box marked AC, which indicates that the output voltage is AC. The top of the box is marked 0–120/208 V–5 A. The 0–120/208 V indicates that the output voltage can be adjusted between 0 and 120 V between any of the 4, 5, or 6 terminals and N, and from 0 to 208 V between any of the 4, 5, or 6 terminals. The voltage is adjusted by the control knob located above these terminals. The markings around the control knob range from 0 to 100. These markings indicate the percentage of the full output voltage. For example, if a voltmeter were connected across terminals 4 and N, it would be seen that the full output voltage is 120 V. If the control knob is adjusted to the 50 position, the voltmeter would indicate a voltage of 60 V, which is 50% of the full output voltage. Now assume that the voltmeter is connected across terminals 4 and 6. The full output voltage of this connection is 208 V. If the control knob were set at the 50 position, the voltmeter would indicate a voltage of 104 V which is 50% of 208 V. The 5 A indicates that the maximum output current of this section of the power supply is 5 amps.

4. Locate the terminals marked 7 and N. Notice that these terminals are labeled DC, which indicates that the output voltage of this section is direct current. This section is marked 0–120 V–8 A. The 0–120 indicates that the voltage can be adjusted between 0 and 120 V. In actual practice, however, this voltage can be adjusted above 120 V, because this DC voltage is provided by a three-phase full-wave bridge rectifier. This type of rectifier has an average voltage value that is close to the peak voltage of the AC wave form producing it. The 8 A indicates that the maximum output current for this section is 8 amps. The output voltage is adjusted by the same control knob that is used to adjust the AC voltage on terminals 4, 5, and 6.

5. Locate the terminals marked 8 and N. The voltage produced by these terminals is DC also. This section is labeled 120 V–2 A. Since there is no zero before the 120 it indicates that this voltage is fixed and cannot be adjusted. In actual practice the output voltage is actually above 120 V. The 2 A indicates that the maximum current for this section is 2 amps. The DC voltage for this section is produced by a three-phase half-wave rectifier.

6. Locate the three pilot lights above the voltage control knob. These pilot lights indicate that all three phases are present when the main switch is turned on.

7. Locate the 120 V, 15 A receptacle below the main power switch. This receptacle is controlled by the main power switch and is used to provide power to external devices such as oscilloscopes, strobe lights, and so on.

8. Locate the voltmeter mounted above the main switch. A rotary switch located to the right of the meter is used to connect the meter to different points on the power supply. For example, if the switch is set for the 4–6 position, the meter will measure the voltage between terminals 4 and 6. If the switch is set for the 7–N position, the meter will measure the voltage between terminals 7 and N. All points on the power supply labeled N are connected together and are the same point electrically.

9. If the power supply is in the cabinet, remove the twist lock plug from the rear of the unit. Release the power supply with the module release tool and remove the unit from the cabinet. Locate the following items:

   A. The three autotransformers used to provide variable AC and DC voltages.

   B. The main power switch. Notice that this switch is actually a 20-A three-phase circuit breaker.

   C. The voltmeter and rotary switch.

   D. Two large filter capacitors.

   Turn the power supply on its side so the bottom can be seen. Locate the following items:

   A. Six individual circuit breakers used to protect different outputs. Notice that all of these circuit breakers can be reset by a common reset button located on the front of the power supply.

   B. The three-phase bridge rectifier. It is characterized by the two heat sinks, each of which contain three press-mounted diodes.

10. Return the power supply to the cabinet and reconnect the twist lock connector.

11. List the voltage and maximum current rating between the terminals listed in the chart shown in Figure 1-2. Indicate whether the voltages are AC or DC and whether they are *variable* (can be adjusted between 0 and 208 or 0 and 120) or *fixed* (have only one voltage value that cannot be adjusted).

12. Connect the circuit shown in Figure 1-3. In this manual, power supply connections will be shown at the left side of the diagram. Figure 1-3 shows that a 250-volt AC meter is connected across terminals 1 and 2 of the power supply. The EMS 8426 AC metering module (AC voltmeter module) contains three AC voltmeters, Figure 1-4. Each voltmeter has three connection terminals. One terminal is labeled (+/–), one is labeled 100, and the third is labeled 250. The (+/–) terminal is the common terminal of the voltmeter. If one lead is connected to the (+/–) terminal and the other lead is connected to the 100 terminal, the AC voltmeter has a full scale value of 100 V. If one lead is connected to the (+/–) terminal and the other lead is connected to the 250 terminal, the AC voltmeter has a full scale value of 250 V. Because this is an AC meter, it is not polarity-sensitive. This means that either meter terminal can be connected to terminal 1 or 2 of the power supply and the meter will give an upscale indication.

Terminals	Voltage (volts)	AC/DC	Fix/Var (F) (V)	Maximum current (amps)
1–2				
1–3				
2–3				
1–N				
2–N				
3–N				
4–5				
4–6				
5–6				
4–N				
5–N				
6–N				
7–N				
8–N				

FIGURE 1-2

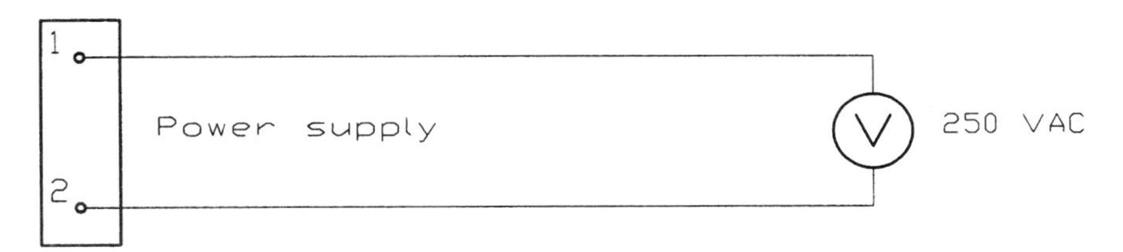

FIGURE 1-3 A 250-VAC voltmeter is connected to the power supply.

13. Turn on the power supply and record the voltage across terminals 1 and 2. Turn the voltage control knob to its full clockwise position while observing the voltmeter. Is this voltage variable? ________________
14. Record the voltage between terminals 1 and 2. ________________ V
15. **Return the voltage control knob to zero and turn off the power supply.**
16. Reverse the leads of the AC voltmeter connected to power supply terminals 1 and 2.
17. Turn on the power supply. Does the voltmeter give an upscale reading or does it read down scale?

    ____
18. **Turn off the power supply.**

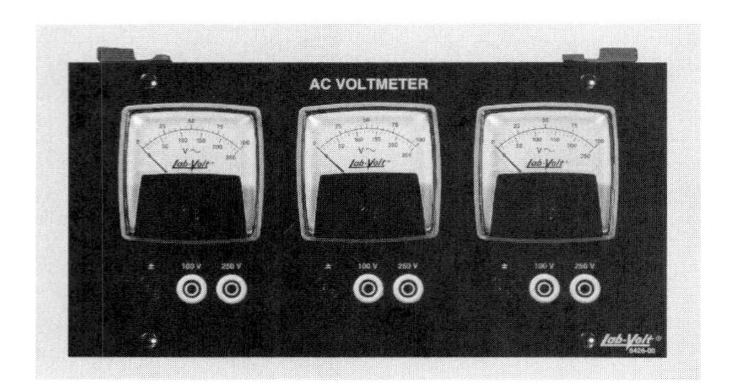

FIGURE 1-4 AC Voltmeter module (Courtesy of Lab-Volt® Systems, Inc.)

19. Measure the voltage between the two terminals in each set listed in the chart shown in Figure 1-5. Indicate whether the voltage is variable or fixed. **Turn off the power supply each time before reconnecting the voltmeter.**
20. Compare these measured values with the listed values.
21. **Turn off the power supply.**
22. Connect the circuit shown in Figure 1-6. Notice that the EMS 8412 DC metering module contains three meters, Figure 1-7. One meter is a voltmeter with two ranges, 20 and 200 V. This meter has a common terminal that is negative. If one lead is connected to the negative terminal and the other lead is connected to the 20 terminal, the voltmeter has a full range value of 20 VDC full scale. If one lead is connected to the negative terminal and the other lead is connected to

Terminals	Voltage (volts)	Fix/Var (F) (V)
1–2		
1–3		
2–3		
1–N		
2–N		
3–N		
4–5		
4–6		
5–6		
4–N		
5–N		
6–N		

FIGURE 1-5

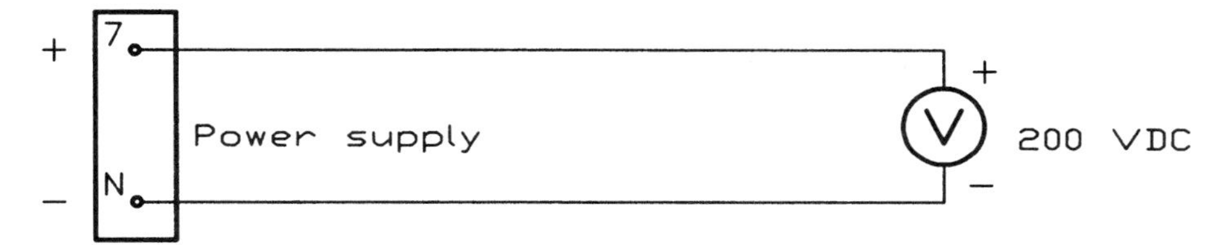

FIGURE 1-6 A DC voltmeter is connected to the power supply.

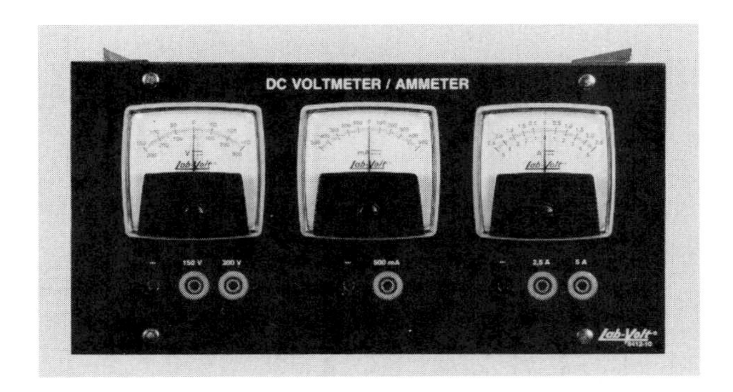

FIGURE 1-7 DC metering module (Courtesy of Lab-Volt® Systems, Inc.)

the 200 terminal, the voltmeter has a range of 200 VDC full scale. Because this is a DC meter, it is polarity-sensitive. If the positive and negative terminals are reversed, the meter will attempt to move down scale.

23. Turn on the power supply and rotate the voltage control knob to the full clockwise position. What is the maximum DC voltage that can be obtained at terminal 7?

    ______________ V

24. Compare this value with the listed value.
25. Is this voltage variable?

    ____

26. **Return the voltage control knob to the zero position and turn off the power supply.**
27. Reverse the voltmeter connections to power supply terminals 7 and N. Turn on the power supply and slowly increase the voltage while observing the DC voltmeter. Does the meter pointer move down scale?

    ____

28. Is the DC voltmeter polarity-sensitive or not polarity-sensitive?

    ____

29. **Return the voltage knob to the zero position and turn off the power supply.**
30. Reconnect the 200-volt DC meter to power supply terminals 8 and N. The positive lead of the voltmeter should be connected to terminal 8.

31. Turn on the power supply and rotate the voltage control knob to the full clockwise position while observing the voltmeter. Record the DC voltage between terminals 8 and N.

    ______________ V

32. Compare this value with the listed value.
33. Is the voltage variable or fixed?

    ____

34. **Return the voltage control knob to the zero position and turn off the power supply.**

    NOTE: The indicated settings of the voltmeters in these experiments are generally set for the highest expected value of voltage to be obtained during the experiment. It may sometimes be helpful to lower the range setting of the voltmeter to obtain a more accurate measurement. There is no harm in lowering the range setting as long as the voltmeter is not over-ranged. If possible, the power should be turned off before changing the voltmeter leads.

AMMETERS

Ammeters, unlike voltmeters, are very-low-impedance devices.

An ammeter must never be connected across a voltage source.

If an ammeter were connected directly to the power supply, it would produce a short circuit and damage the ammeter. An ammeter must always be connected in series with some type of current-limiting device. The ammeter in Figure 1-8, for example, is connected in series with a 300-Ω resistor. The 300-Ω resistor is used to limit the flow of current in the circuit.

35. Using the EMS 8311 resistance load module, Figure 1-9 the EMS 8425 AC metering module (AC ammeter module), Figure 1-10 and the EMS 8426 AC metering module (AC voltmeter module), connect the circuit shown in Figure 1-5. Notice that the ammeter, like the voltmeter contains more than one range setting. The (+/–) terminal is the common terminal of the ammeter. Different ammeter ranges are obtained by connecting one lead to the (+/–) terminal and the other lead to the terminal indicating the desired range.
36. Assume that a voltage of 120 V is to be applied to the circuit. Compute the amount of current that should flow using the formula

$$I = \frac{E}{R}$$

I = ______________

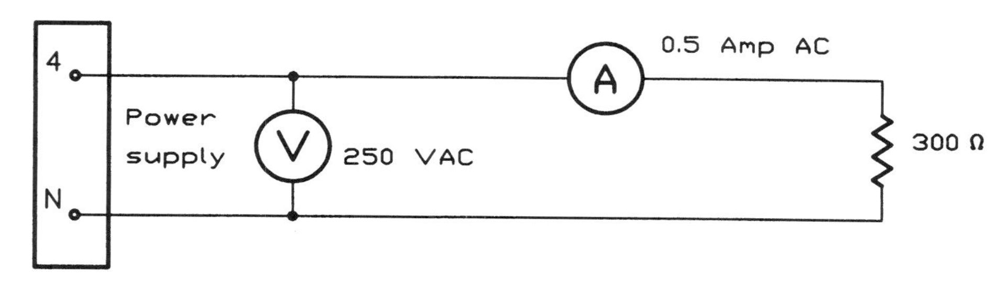

FIGURE 1-8 Measuring voltage and current

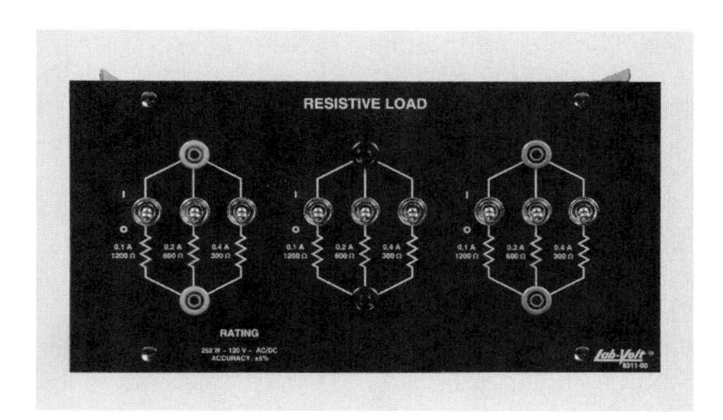

FIGURE 1-9 Resistive load module (Courtesy of Lab-Volt® Systems, Inc.)

FIGURE 1-10 AC ammeter module (Courtesy of Lab-Volt® Systems, Inc.)

37. Turn on the power supply and adjust the output voltage for a value of 120 V. Measure the amount of current flow through the circuit.

    ______________ A

38. **Return the voltage to zero and turn off the power supply.**
39. Disconnect the circuit and return the components to their proper place.

Review Questions

1. What is the maximum current rating of terminals 1, 2, and 3?

   ______________ A

2. What is the maximum voltage that can be obtained across terminals 4, 5, and 6?

   ______________ V

3. Is the voltage across terminals 4, 5, and 6 variable or fixed?

   ____

4. Is the voltage across terminals 1 and N variable or fixed?

   ____

5. What voltage is produced across terminals 3 and N?

   ______________ V

6. What is the maximum current rating of terminals 8 and N?

   ______________ A

7. Is the voltage produced across terminals 7 and N AC or DC?

   ____

8. What is the maximum output current of terminals 4, 5, and 6?

   ______________ A

9. Can the voltmeter be set to measure the voltage across terminals 4 and 5?

   ____

10. Can the voltmeter be set to measure the voltage across terminals 3 and N?

   ____

Exercise 2

Ohm's Law

Objectives

After completing this lab you should be able to:

- Discuss the relationship of voltage, current, resistance, and power in an electrical circuit.
- Compute the values of voltage, current, resistance, and power using Ohm's law formulas.

Materials and Equipment

Power supply module	EMS 8821
DC metering module	EMS 8412
Variable-resistance module	EMS 8311

Discussion

All values of voltage, current, resistance, and power in an electrical circuit originating from a single power source can be computed using Ohm's law. A set of mathematical formulas are used to determine these different quantities.

Different letters are used to represent electrical quantities. For example, the letter *V* (for volts) or *E* (for EMF, electromotive force) are generally used to represent the value for voltage in an equation. The letter *I* (for intensity of current) or *A* (for amps) is used to represent the value of current. The letter *R* is used to represent the value for resistance (measured in ohms, **Ω**, and the letter *W* (for watts) or *P* (for power) are used to represent the values of true power, or watts.

The following formulas can be used for finding values of voltage, current, resistance, or power if two other quantities are known.

The formulas for finding voltage are

$$E = I \times R \qquad E = \frac{P}{I} \qquad E = \sqrt{P \times R}$$

The formulas for finding current are

$$I = \frac{E}{R} \qquad I = \frac{P}{E} \qquad I = \sqrt{\frac{P}{R}}$$

The formulas for finding resistance are

$$R = \frac{E}{I} \qquad R = \frac{E^2}{P} \qquad R = \frac{P}{I^2}$$

The formulas for finding power are

$$P = E \times I \qquad P = I^2 \times R \qquad P = \frac{E^2}{R}$$

Name ______________________________ Date ______________

Procedure

1. Connect the circuit shown in Figure 2-1.
2. Using the resistance chart shown in Figure 2-2, set the variable resistance module for a resistance of 100 Ω.

   The chart has been divided into three sections: first section, second section, and third section. These sections correspond to the three different sections on the load module. Each of the three sections on the load module contains three resistors: 1200 OHM, 600 OHM, and 300 OHM. These three resistors can be connected separately or in parallel with others by closing (turning on) the switch that controls that particular resistor. To obtain the 100 OHM value, connect all three sections of the load module together as shown in Figure 2-1. Find 100 in the far left-hand column of the chart. This column lists the total value of resistance that will be obtained when the proper resistors have been connected in parallel. Next, find the resistance value that should be turned on for each section of the load module. In this example, a value of 300 OHMs is indicated for each of the three sections of the load module. Close (turn on) the switch that controls each of the 300-OHM resistors for each section.
3. Assume an applied voltage of 120 V and compute the amount of current flow in this circuit using the formula

$$I = \frac{E}{R}$$

   I = ______________ A
4. Compute the amount of power used in this circuit by using the formula

$$P = \frac{E^2}{R}$$

   P = ______________ W
5. Turn on the power supply and adjust the voltage for a value of 120 V.
6. Measure the amount of current flow in the circuit and compare this value with the computed value.

   I = ______________ A
7. **Return the voltage to zero and turn off the power supply.**
8. Reset the resistance module to a value of 60 Ω.

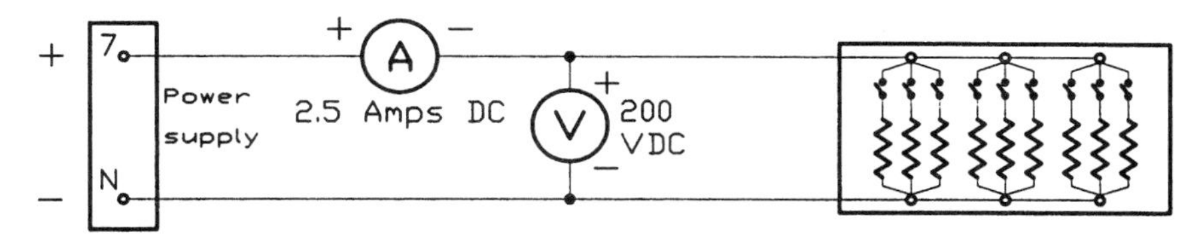

FIGURE 2-1

Resistances obtainable for variable-resistance module (All sections connected in parallel)			
Total resistance (ohms)	First section Closed switches (ohms)	Second section Closed switches (ohms)	Third section Closed switches (ohms)
1200	1200	None	None
600	600	None	None
400	1200 & 600	None	None
300	300	None	None
240	1200 & 300	None	None
200	600 & 300	None	None
171.4	1200 & 600 & 300	None	None
150	1200 & 600 & 300	1200	None
133.3	1200 & 600 & 300	600	None
120	600 & 300	300	None
109.1	1200 & 600 & 300	300	None
100	300	300	300
92.3	1200 & 600 & 300	600 & 300	None
85.7	1200 & 600 & 300	1200 & 600 & 300	None
80	1200 & 600 & 300	1200 & 600 & 300	1200
75	1200 & 600 & 300	1200 & 600 & 300	600
70.6	1200 & 600 & 300	600 & 300	300
66.7	1200 & 600 & 300	1200 & 600 & 300	300
63.1	1200 & 600 & 300	1200 & 600 & 300	1200 & 300
60	1200 & 600 & 300	1200 & 600 & 300	600 & 300
57.1	1200 & 600 & 300	1200 & 600 & 300	1200 & 600 & 300

FIGURE 2-2 Other values of resistance can be obtained with different parallel combinations.

9. Compute the amount of current flow in this circuit.

   I = ______________ A

10. Compute the amount of power in this circuit.

    P = ______________ W

11. Turn on the power supply and adjust the voltage for a value of 120 V.

12. Measure the amount of current flow through the circuit and compare this with the computed value.

    I = ______________ A

13. **Return the voltage to zero and turn off the power supply. Disconnect the circuit and return the components to their proper place.**

Review Questions

1. An electric iron has a resistance of 16 Ω. How much current will flow if the iron is connected to a 120-V circuit?

   I = ______________ A

2. How much power is used by the iron in question 1?

   P = ______________ W

3. A light bulb has a rating of 100 W when connected to 120 V. How much current will flow through this light bulb?

   I = ______________ A

4. What is the resistance of the filament of the light bulb in question 3. Assume the bulb is in operation.

   R = ______________ Ω

5. An electric range element has a resistance of 24 Ω and produces 2400 W of heat when operating. How much current will flow through this element?

   I = ______________ A

Exercise 3

Series Circuits

Objectives

After completing this lab you should be able to:

- Discuss the properties of series circuits.
- Compute values of voltage, current, and resistance for series circuits.
- List rules for solving electrical values of series circuits.

Materials and Equipment

Power supply module	EMS 8821
AC voltmeter module	EMS 8426
AC ammeter module	EMS 8425
Variable-resistance module	EMS 8311
Ohmmeter (supplied by student)	

Discussion

A series circuit is a circuit that has only one path for current flow. Because there is only one path for current flow, the current is the same at all points in the circuit. There are three rules that can be used with Ohm's law for finding values of voltage, current, resistance, and power in any series circuit. These rules are:

1. The current is the same at all points in the circuit.
2. The total resistance is the sum of the resistances of the individual resistors.
3. The applied voltage is equal to the sum of the voltage drops across all the individual resistors.

CONNECTING THE CIRCUIT

Students often become perplexed when first attempting to connect an electrical circuit. Although a circuit can appear simple in the form of a schematic diagram, it may become confusing when an attempt is made to actually connect the circuit. The main reason for this is that when a circuit is connected it generally bears little resemblance to the schematic diagram. The circuit shown in Figure 3-1 is a good example. In this circuit, three resistors are connected in series. A 0.5-A ammeter is connected in the circuit and a voltmeter with a full scale range of 100 V is connected across each resistor. The entire circuit is connected to terminals 4 and 5 of the power supply.

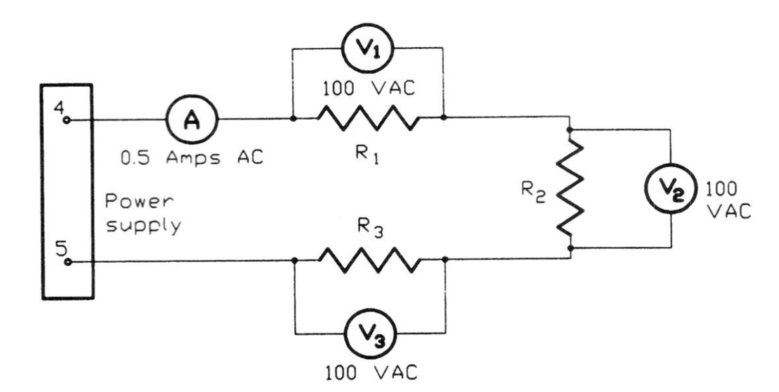

FIGURE 3-1 Basic series circuit

The first step will be to number the circuit in Figure 3-1. This is done by starting at one end of the circuit with the number 1. Each device connected to this point will be labeled 1 also. Each time a device is encountered, the number will change. A set of numbers can never be used more than once. In Figure 3-2, numbers have been added to the circuit shown in Figure 3-1. A 1 was placed beside power supply terminal 4 and beside one side of the 0.5-A AC ammeter. Notice that there is no component between these two places. The 0.5-A AC meter is a component. When going through a component, the number set must change. The other side of the ammeter is labeled with the number 2. There is also a 2 placed beside one side of the 100-VAC voltmeter 1 (V_1) and one side of resistor 1 (R_1). The other side of voltmeter 1 is labeled with a 3, the other side of resistor 1 is labeled with a 3 also. One side of voltmeter 2 is labeled with a 3, and one side of resistor 2 is labeled with a 3. Notice that all these points are electrically the same. There is no break in the circuit between any of these components.

The other side of voltmeter 2 is labeled with a 4; the other side of resistor 2 is labeled with a 4. One side of voltmeter 3 is labeled with a 4, and one side of resistor 3 is labeled with a 4.

The other side of voltmeter 3 is labeled with a 5; the other side of resistor 3 is labeled with a 5, and terminal 5 of the power supply is labeled with a 5.

Now that the schematic has been numbered, the actual components that are to be used in the circuit will be labeled with the same numbers. Figure 3-3 shows the components that will be used to connect the circuit. Terminals 4 and 5 of the power

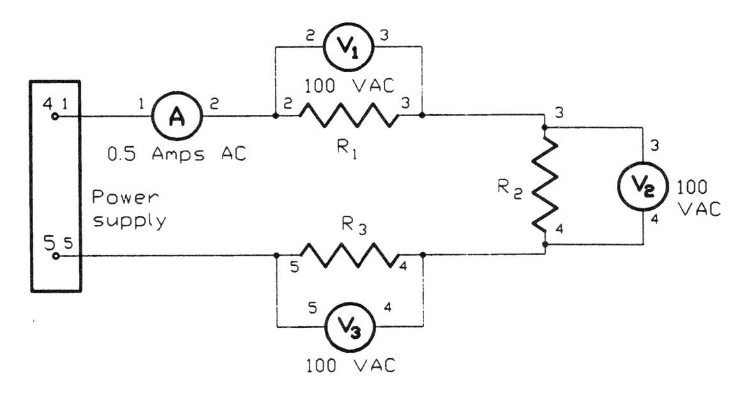

FIGURE 3-2 Numbers are placed on the schematic.

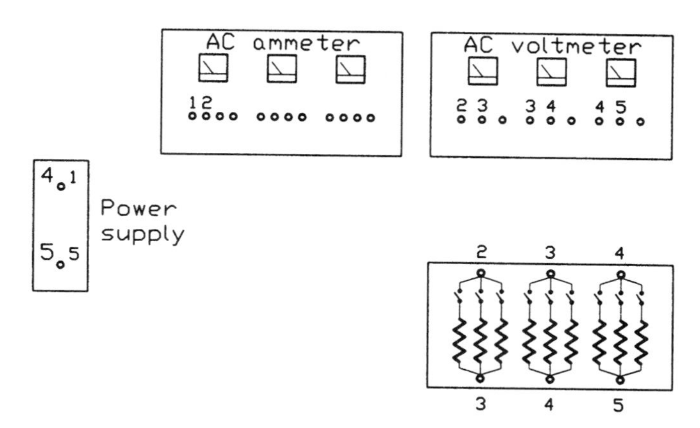

FIGURE 3-3 Numbers are placed on the components.

supply, the AC ammeter module, the AC voltmeter module, and the variable-resistance load module are shown. Notice that numbers have been placed above or below certain terminals on the modules. A 1 has been placed beside terminal 4 of the power supply, and above the common terminal of one of the AC ammeters. The schematic in Figure 3-2 shows that a 1 has been placed beside those same two components. A 2 has been placed above the 0.5 terminal of the AC ammeter, the common terminal of voltmeter 1, and one side of resistor 1. A 3 has been placed above the 100-V range terminal of voltmeter 1, below the second side of resistor 1, and above the common terminal of voltmeter 2 and one side of resistor 2. Notice that the schematic has a 3 placed beside the same terminals.

A 4 has been placed above the 100-V terminal of voltmeter 2, below the second side of resistor 2, and above the common terminal of voltmeter 3 and one side of resistor 3. Check these numbers against the numbers shown on the schematic in Figure 3-2.

A 5 has been placed above the 100-V terminal of voltmeter 3, below the second side of resistor 3, and at terminal 5 of the power supply. This completes the numbering of the components.

Now that the components have been numbered to correspond with the numbers on the schematic, the circuit is connected by connecting all like numbers together. All the 1s are connected together, all the 2s are connected together, all the 3s are connected together, and so on. This numbering system can be used with any schematic regardless of how involved or complicated it is, if the rules are followed.

Variable Voltage
Source so that
Volts start at zero
then slowly turn up
making sure all connections
are correct. especially current
meter

Name ______________________________ Date ______________

Procedure

1. Connect the circuit shown in Figure 3-4.
2. Set the variable resistance module so that the resistors in section 1 have a total resistance of 300 Ω, the resistors in section 2 have a total resistance of 200 Ω, and the resistors in section 3 have a total resistance of 240 Ω.

   NOTE: Use the chart in Figure 2-2.
3. Compute the value of total resistance (R_T) in the circuit, using the formula

$$R_T = R_1 + R_2 + R_3$$

   R_T = ______________ Ω
4. Disconnect the circuit from the power supply and measure the total resistance of the circuit with an ohmmeter. Compare this value with the computed value.

   R_T = ______________ Ω
5. Reconnect the circuit to terminals 4 and 5 of the power supply.
6. If an applied voltage of 208 V is assumed for this circuit, compute the total amount of current flow (I_T) using the formula

$$I_T = \frac{E_T}{R_T}$$

   I_T = ______________ A
7. Now that the current flow through the circuit is known, the voltage drop across each resistor can be computed using the following formulas

$$E_1 = I_1 \times R_1$$

   E_1 = ______________ V

$$E_2 = I_2 \times R_2$$

   E_2 = ______________ V

$$E_3 = I_3 \times R_3$$

   E_3 = ______________ V

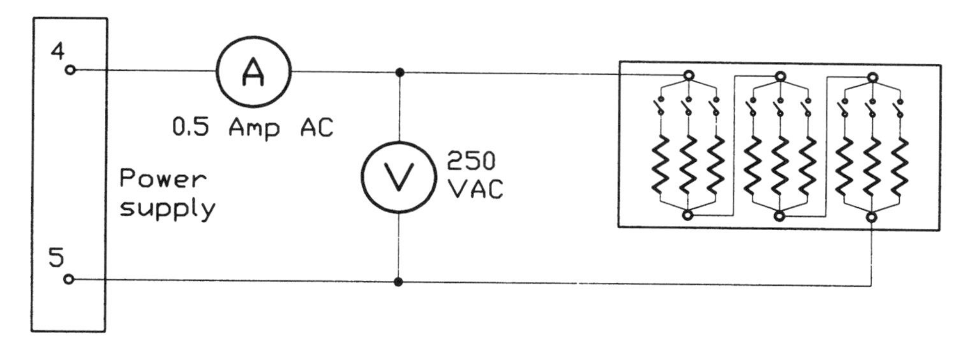

FIGURE 3-4 Each section of the resistance module is connected in series.

8. Turn on the power supply and adjust the output voltage for a value of 208 V.
9. Measure the amount of current flow in the circuit and compare this value with the computed value.

   I_T = ______________ A

10. **Turn off the power supply and connect a 100-VAC voltmeter across the first section of resistance.** This should be the section set for 300 Ω.
11. Turn on the power supply and measure the voltage drop across this section of resistance. Compare this value with the computed value.

    E_1 = ______________ V

12. **Turn off the power supply and reconnect the 100-VAC voltmeter across the second section of resistance.**
13. Turn on the power supply and measure the voltage drop across this section. Compare this value with the computed value.

    E_2 = ______________ V

14. **Turn off the power supply and reconnect the 100-VAC voltmeter across the third section of resistance.**
15. Turn on the power supply and measure the voltage drop across this section of resistance. Compare this value with the computed value.

    E_3 = ______________ V

16. **Return the voltage control knob to the zero volt position and turn off the power supply. Disconnect the circuit and return the components to their proper place.**

Review Questions

1. Define a series circuit.

   ____

2. State the three rules for series circuits.

   A. ____

   B. ____

   C. ____

3. A series circuit has resistance values of 160 Ω, 100 Ω, 82 Ω, and 120 Ω. What is the total resistance of this circuit?

   ______________ Ω

4. If a voltage of 24 V is applied to this circuit, what will be the total amount of current flow in the circuit?

   ______________ A

5. How much voltage will be dropped across each of the resistors?

   160______________ V

   100______________ V

   82______________ V

   120______________ V

Exercise 4

Parallel Circuits

Objectives

After completing this lab you should be able to:

- Compute the values in a parallel circuit using rules for parallel circuits and Ohm's law.
- Connect a parallel circuit and measure electrical values using instruments.

Materials and Equipment

Power supply module	EMS 8821
Variable-resistance module	EMS 8311
AC ammeter module	EMS 8425
Single-phase wattmeter module	EMS 8431
AC voltmeter module	EMS 8426
Ohmmeter (supplied by student)	

Discussion

A parallel circuit is a circuit that has more than one path for current flow. There are three rules that can be used for finding the unknown values in a parallel circuit. These rules are:

1. The voltage is the same across all the components in a parallel circuit (voltage remains the same).
2. The total current is the sum of the currents flowing through the individual paths (current adds).
3. The reciprocal of the total resistance is equal to the sum of the reciprocals of the individual resistors.

When these three rules are used in conjunction with Ohm's law, the values of voltage, current, resistance, and power can be computed for a parallel circuit. Figure 4-1 shows a schematic diagram of a parallel circuit. Notice that the output leads of the power supply, in this case a generator, are connected to each resistor. The same voltage is, therefore, applied to each one. Ammeters have been connected to each of the resistors, and one ammeter has been connected in the circuit at the output of the generator. The ammeter labeled A_T will indicate the total current in the circuit. The other ammeters will indicate the amount of current flow through each individual path or branch. Because the current for each resistor must flow through ammeter A_T, it will indicate the sum of the currents flowing through the three resistors.

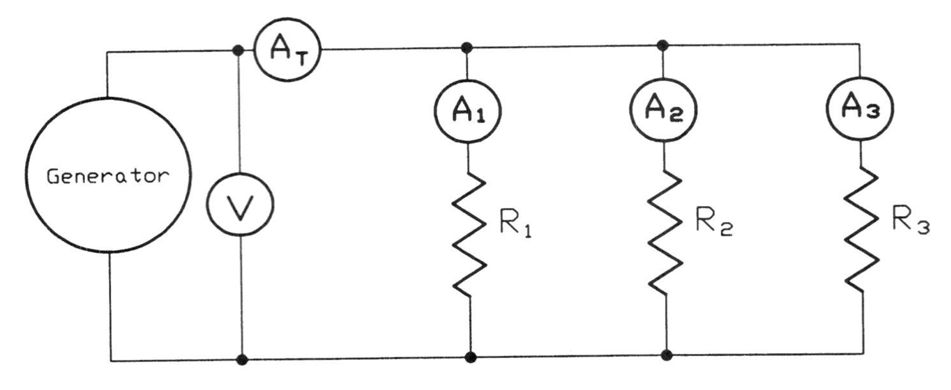

FIGURE 4-1 Basic parallel circuit

Name ______________________________ Date ______________

Procedure

1. Connect the circuit shown in Figure 4-2. If necessary, connect the wattmeter module shown in Figure 4-3 to the 24-volt supply located on the power supply module. (Some wattmeter modules do not require this connection.)

2. Set the resistance of the first set of resistors for a value of 240 Ω, the second set of resistors for a value of 400 Ω, and the third set of resistors for a value of 300 Ω.

3. Find the total resistance of the circuit, using the formula

$$\frac{1}{R_T} = \frac{1}{R_1} + \frac{1}{R_2} + \frac{1}{R_3}$$

   R_T = ______________ Ω

4. Disconnect the circuit from wattmeter terminals 3 and 4.

5. Using an ohmmeter, measure the total resistance of the circuit.

   R_T = ______________ Ω

6. Compare this measured value with the computed value in step 3.

7. Reconnect the circuit to terminals 3 and 4 of the wattmeter.

8. Because the resistors in this circuit are connected in parallel with each other, the same voltage will be connected across all the resistors. The value of this voltage will be the applied voltage of 120 VAC. Using the formula shown below, compute the amount of current flow through each resistor.

$$I = \frac{E}{R}$$

   I_1 = ______________ A I_2 = ______________ A I_3 = ______________ A

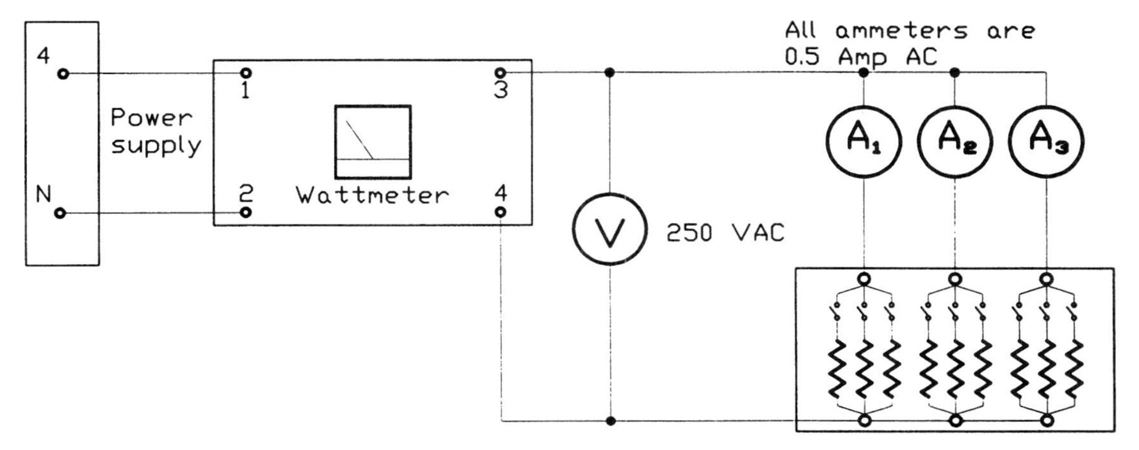

FIGURE 4-2 Parallel circuit connected using the Lab-Volt® EMS modules

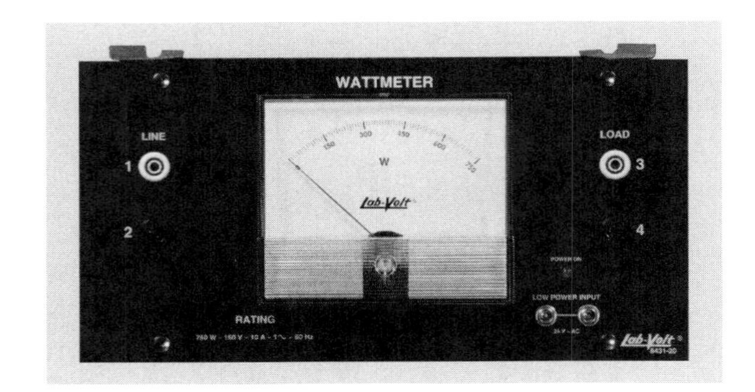

FIGURE 4-3 Single-phase wattmeter (Courtesy of Lab-Volt® Systems, Inc.)

9. Add the values of the currents through all three resistors to find the amount of total current flow in the circuit.

$$I_T = I_1 + I_2 + I_3$$

I_T = ______________ A

10. Find the total amount of power consumed in this circuit, using the formula below. Be sure to use the total values of voltage and current.

$$P = E \times I$$

P_T = ______________ W

11. Turn on the power supply and adjust the voltage for a value of 120 V.

12. Measure the amount of current flow through each resistor.

NOTE: In step 8, the value of I_1 was calculated at 0.5 A. The variable-resistance module has an accuracy of +/– 5%. For this reason it may be necessary to change the range of ammeter A_1 to a full scale value of 2.5 A.

I_1 = ______________ A I_2 = ______________ A I_3 = ______________ A

13. Compare these values with the computed values.

14. Measure the amount of total power used in this circuit.

P_T = ______________ W

15. Compare this value with the computed value.

16. **Return the voltage to 0 V and turn off the power supply.**

17. Readjust the values of resistance on the variable-resistance module so that resistor R_1 has a value of 600 Ω, resistor R_2 has a value of 1200 Ω, and resistor R_3 has a value of 400 Ω.

18. Compute the total resistance of this circuit using the formula

$$R_T = \frac{R_1 \times R_2}{R_1 + R_2}$$

R_T = ______________ Ω

19. Disconnect the circuit from terminals 3 and 4 on the wattmeter module.

20. Using an ohmmeter, measure the amount of total resistance in the circuit.

    R_T = ______________ Ω

21. Compare the measured value with the computed value.

22. Reconnect the circuit to terminals 3 and 4 on the wattmeter module.

23. Using Ohm's law, compute the value of current flow through each resistor if 120 V is applied to the circuit.

    I_1 = ______________ A I_2 = ______________ A I_3 = ______________ A

24. Compute the amount of total current in the circuit using the formula

$$I_T = I_1 + I_2 + I_3$$

    I_T = ______________ A

25. Compute the amount of total power used in the circuit using the formula

$$P = E \times I$$

    P_T = ______________ W

26. Turn on the power supply and adjust the voltage for a value of 120 VAC.

27. Measure the amount of current flow through each resistor.

    I_1 = ______________ A I_2 = ______________ A I_3 = ______________ A

28. Compare these values with the computed values.

29. Measure the amount of total power used in this circuit.

    P_T = ______________ W

30. Compare this value with the computed value.

31. **Return the voltage control to 0 V and turn off the power supply.**

32. Disconnect the circuit and replace the equipment in its proper place.

Review Questions

1. A parallel circuit has resistors that have values of 2200 Ω, 3600 Ω, 4700 Ω, and 3300 Ω. What is the total resistance of the circuit?

   R_T = ______________ Ω

2. A parallel circuit has five branches. The current flow through each branch is as follows: 2.4 A, 0.98 A, 1.2 A, 4.3 A, and 0.88 A. What is the total current flow in the circuit?

   I_T = ______________ A

3. Four 100-W lamps are connected in parallel to 120 VAC. What is the total power being consumed by the circuit, and what is the total current flow?

   I_T = ______________ A P_T = ______________ W

4. Fours resistors having a value of 100 Ω each are connected in parallel. If another 100-Ω resistor is connected in parallel with the existing four, will the total resistance of the circuit increase or decrease?

   If these five resistors are connected to 240 V, what will be the total current flow in the circuit?

   I_T = ______________ A

5. A circuit has three resistors connected in parallel. The total resistance of the circuit is 80 Ω. Two of the resistors have values of 400 Ω and 600 Ω. What is the value of the third resistor?

   NOTE: The reciprocal of the total resistance of a parallel circuit is the sum of the reciprocals of the individual resistors. Therefore, the reciprocal of the third resistor can be found by subtracting the sum of the reciprocals of the two known resistors from the reciprocal of the total resistance.

$$\frac{1}{R_3} = \frac{1}{R_T} - \left(\frac{1}{R_1} + \frac{1}{R_2}\right)$$

or

$$\frac{1}{R_3} = \frac{1}{R_T} - \frac{1}{R_1} - \frac{1}{R_2}$$

   R_3 = ______________

14ωω

Exercise 5

Combination Circuits

Objectives

After completing this lab you should be able to:

- Compute the values in a combination circuit using rules for series and parallel circuits and Ohm's law.
- Connect a combination circuit and measure electrical values using instruments.

Materials and Equipment

Power supply	EMS 8821
Variable-resistance module	EMS 8311
AC ammeter module	EMS 8425
AC voltmeter module	EMS 8426
Ohmmeter (supplied by student)	

Discussion

Combination circuits contain elements of both series and parallel circuits. When computing the values of voltage, amperage, resistance, and power for a combination circuit, you must first determine which parts of the circuit are connected in series and which are connected in parallel. To do this, you generally have to trace the path of current flow through the circuit. Recall that a series circuit is a circuit that has only one path for current flow and a parallel circuit has more than one path for current flow.

Once it has been determined which parts of a circuit are connected in series and which are connected in parallel, the rules for series and parallel circuits can be used with Ohm's law to determine the electrical values in the circuit. The rules for series and parallel circuits follow.

SERIES CIRCUITS

1. The sum of the voltage drops across individual resistors is equal to the applied voltage (voltage drops add).
2. The current is the same at all points in a series circuit (current remains the same).
3. Total resistance is equal to the sum of all the individual resistors (resistance adds).

PARALLEL CIRCUITS

1. The voltage is the same across all the components in a parallel circuit (voltage remains the same).
2. The total current is the sum of the currents flowing through the individual paths (current adds).
3. The reciprocal of the total resistance is equal to the sum of the reciprocals of all the individual resistors.

Name ______________________ Date ____________

Procedure

1. Connect the circuit shown in Figure 5-1.
2. Determine which parts of the circuit are connected in parallel and which are connected in series by tracing the flow of current through the circuit. To do this, assume that current will flow from terminal 4 of the power supply to terminal 6. As you trace the path, you will see that all the circuit current must flow through resistor R_1. Resistor R_1, therefore, is connected in series with any other components in the circuit.

   After the current leaves resistor R_1, it flows to a junction of resistors R_2 and R_3. Part of the circuit current will flow through resistor R_2, and part will flow through resistor R_3. Because these two resistors provide more than one path for current flow, they are connected in parallel with each other.
3. Compute the values of total circuit resistance and current. The first step in finding the total values of current and resistance in this circuit is to simplify the circuit. To do this, reduce the parallel combination of resistors R_2 and R_3 to a single resistor value by computing the total resistance of these two resistors. This value will be known as R_C (resistance of the combination).

$$R_C = \frac{R_2 \times R_3}{R_2 + R_3}$$

$$R_C = \frac{400\ \Omega \times 600\ \Omega}{400\ \Omega + 600\ \Omega}$$

$$R_C = \frac{240,000\ \Omega}{1000\ \Omega}$$

$$R_C = 240\ \Omega$$

or

$$R_C = \frac{1}{\frac{1}{R_2} + \frac{1}{R_3}}$$

$$R_C = \frac{1}{\frac{1}{400} + \frac{1}{600}}$$

$$R_C = \frac{1}{0.004166}$$

$$R_C = 240\ \Omega$$

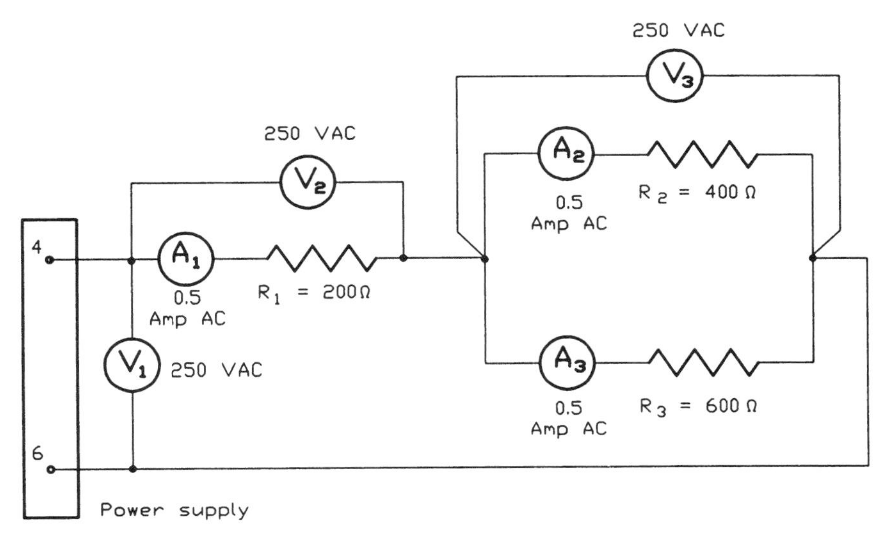

FIGURE 5-1 Combination circuit 1

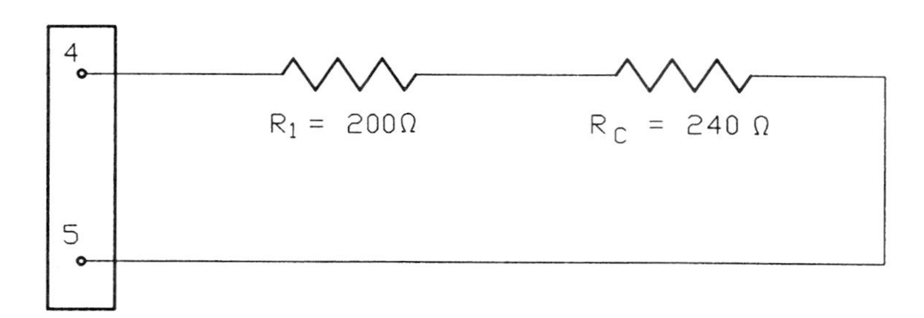

FIGURE 5-2 Simplifying the circuit

Now that the total resistance of the two resistors connected in parallel is known, the circuit can be redrawn as shown in Figure 5-2. The circuit has now become a simple series circuit containing two resistors. The total resistance of the circuit can now be found by adding the two values together.

$$R_T = R_1 + R_C$$
$$R_T = 200\ \Omega + 240\ \Omega$$
$$R_T = 440\ \Omega$$

If a voltage of 200 VAC is applied to the circuit, the total current flow in the circuit can be determined by using Ohm's law.

$$I = \frac{E}{R}$$
$$I = \frac{240\ V}{440\ \Omega}$$
$$I = 0.454\ A$$

4. Determine the voltage drop across each resistor. The equivalent circuit shown in Figure 5-2 is a series circuit. The current flow is, therefore, the same through both resistors. Because the ohmic value of each resistor is known, and the amount of current flowing through each is known, the voltage drop can be computed using Ohm's law.

$$E_1 = I_1 \times R_1$$
$$E_1 = 0.454\ A \times 240\ \Omega$$
$$E_1 = 90.8\ V$$

$$E_C = I_C \times R_C$$
$$E_C = 0.454\ A \times 240\ \Omega$$
$$E_C = 109\ V$$

The values of voltage and current found for the circuit in Figure 5-2 can be applied to the original circuit. Resistor R_C is actually the combination of resistors R_2 and R_3, so the values of voltage and current for resistor R_C apply to the parallel combination. One of the rules for a parallel circuit states that the voltage is the same across all the circuit components. Resistors R_2 and R_3, therefore, have the same voltage drop. This voltage drop is the same as the voltage drop for resistor R_C (109 V).

5. Determine the current flow through each resistor. Because resistor R_1 is connected in series with anything in the circuit, the current flow through resistor R_1 is the same as the total current of the circuit (0.454 A). Now that the voltage drop across resistors R_2 and R_3 is known, the amount of current flow through each can be determined using Ohm's law.

$$I_2 = \frac{E_2}{R_2}$$

$$I_2 = \frac{109\ \text{V}}{400\ \Omega}$$

$$I_2 = 0.272\ \text{A}$$

$$I_3 = \frac{E_3}{R_3}$$

$$I_3 = \frac{109\ \text{V}}{600\ \Omega}$$

$$I_3 = 0.182\ \text{A}$$

NOTE: One of the rules for parallel circuits states that the total current is equal to the sum of the currents flowing through the individual paths. The current flow through resistors R_2 and R_3 should add to equal the total circuit current.

Circuit summary: The values for the circuit shown in Figure 5-1 are as follows:

E_T = 200 VAC	E_1 = 90.8 VAC	E_2 = 109 VAC	E_3 = 109 VAC
I_T = 0.454 A	I_1 = 0.454 A	I_2 = 0.272 A	I_3 = 0.182 A
R_T = 440 Ω	R_1 = 200 Ω	R_2 = 400 Ω	R_3 = 600 Ω

6. Set the resistance values as indicated in the above problem.
7. Disconnect the circuit from terminals 4 and 6 on the power supply.
8. Using an ohmmeter, measure the total circuit resistance.

   R_T = ______________ Ω
9. Compare this value with the computed value for total resistance.
10. Reconnect the circuit to terminals 4 and 6 of the power supply.
11. Turn on the power and adjust the voltage until voltmeter V_1 indicates a voltage of 200 V.
12. Measure the current flow through each resistor using ammeters A_1, A_2, and A_3.

    A_1 ______________ A A_2 ______________ A A_3 ______________ A
13. Compare these measured values with the computed values.
14. Measure the voltage drop across each resistor with voltmeters V_1 and V_2.

    V_1 ______________ V V_2 ______________ V
15. Compare these measured values with the computed values.
16. **Return the voltage to 0 V and turn off the power supply.**

17. Using the variable-resistance module, change the values of resistance in the circuit so that resistor R_1 has a value of 300 Ω, resistor R_2 has a value of 300 Ω, and resistor R_3 has a value of 600 Ω.

18. Compute the combined resistance of resistors R_2 and R_3 using the formula

$$R_C = \frac{R_2 \times R_3}{R_2 + R_3}$$

R_C = ______________ Ω

19. Compute the total resistance of the circuit by adding the resistance of resistors R_1 and R_C.

$$R_T = R_1 + R_C$$

R_T = ______________ Ω

20. Disconnect the circuit from terminals 4 and 6 of the power supply and measure the total resistance of the circuit with an ohmmeter.

R_T = ______________ Ω

21. Compare the measured value with the computed value.

22. Reconnect the circuit to terminals 4 and 6 of the power supply.

23. Assume that a voltage of 200 V is to be applied to the circuit and compute the total current flow in the circuit using the formula

$$I_T = \frac{E_T}{R_T}$$

I_T = ______________ A

24. Compute the voltage drop across each resistor using the formulas

$$E_1 = I_1 \times R_1$$

E_1 = ______________ V

$$E_C = I_C \times R_C$$

E_C = ______________ V

25. Because the voltage dropped across RC is the same as the voltage drop across resistors R_2 and R_3, we can compute the current flow through resistors R_2 and R_3 using the formulas

$$I_2 = \frac{E_2}{R_2}$$

I_2 = ______________ A

$$I_3 = \frac{E_3}{R_3}$$

I_3 = ______________ A

26. Turn on the power supply until voltmeter V_T indicates a value of 200 V.

27. Measure the value of current flow through ammeters A_1, A_2, and A_3.

A_1 ______________ A A_2 ______________ A A_3 ______________ A

28. Compare these measured values with the computed values.
29. Measure the amount of voltage dropped across the resistors with voltmeters V_1 and V_2.

    V_1 ____________ V V_2 ____________ V
30. Compare the measured values with the computed values.
31. **Return the voltage to 0 V and turn off the power supply.**
32. Connect the circuit shown in Figure 5-3.
33. Determine which parts of the circuit are connected in series and which are connected in parallel. To do this, trace the flow of current through the circuit. Assume that current flows from terminal 4 to terminal N of the power supply. One path for circuit current is from terminal 4, through ammeter A_1 to the first circuit branch. Current can then flow through ammeter A_2 to resistor R_1 and then back to terminal N. Current can also flow from ammeter A_1 to the second branch of the circuit, through resistor R_2, ammeter A_3, and resistor R_3. The current can then return to terminal N of the power supply. Because the same current must flow through resistors R_2 and R_3, they are connected in series with each other. Resistor R_1, however, is connected in parallel with resistors R_2 and R_3.
34. Determine the total resistance of the circuit. The first step is to simplify the circuit. Because resistors R_2 and R_3 are connected in series, their ohmic values can be added and the total can be considered as one resistor.

$$R_C = R_2 + R_3$$
$$R_C = 200\ \Omega + 300\ \Omega$$
$$R_C = 500\ \Omega$$

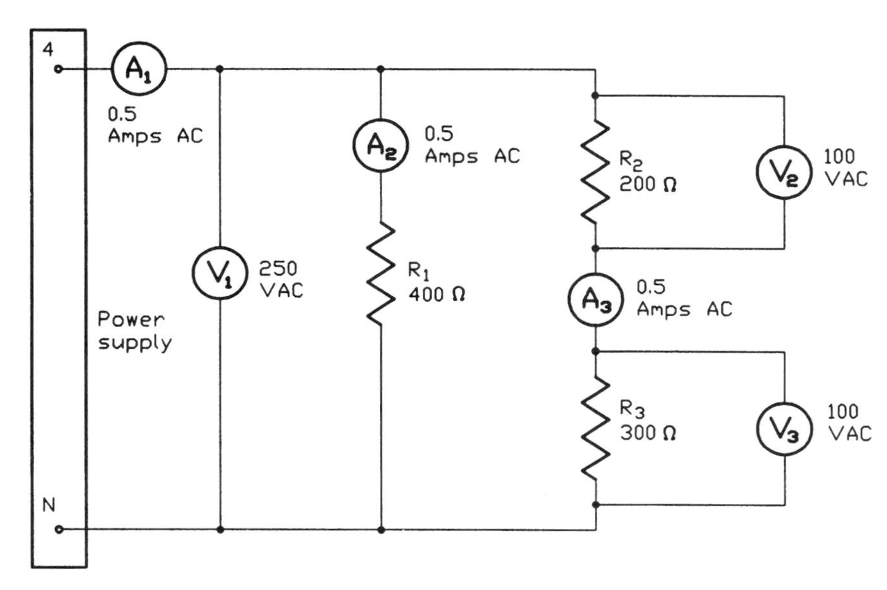

FIGURE 5-3 Combination circuit 2

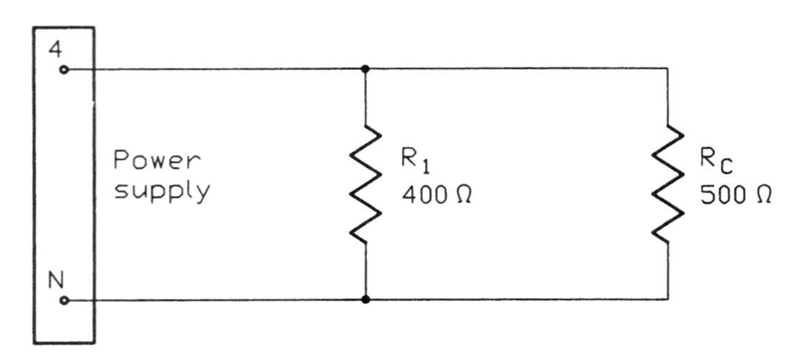

FIGURE 5-4 Simplifying circuit 2

The circuit can now be redrawn to form a simple parallel circuit as shown in Figure 5-4. The total resistance of the circuit can now be computed using the formula

$$R_T = \frac{R_1 \times R_C}{R_1 + R_C}$$

$$R_T = \frac{400\ \Omega \times 500\ \Omega}{400\ \Omega + 500\ \Omega}$$

$$R_T = \frac{200,000\ \Omega}{900\ \Omega}$$

$$R_T = 222.2\ \Omega$$

or

$$R_T = \frac{1}{\frac{1}{R_1} + \frac{1}{R_C}}$$

$$R_T = \frac{1}{\frac{1}{400} + \frac{1}{500}}$$

$$R_T = \frac{1}{0.0045}$$

$$R_T = 222.2\ \Omega$$

35. The total circuit can now be computed using the formula

$$I_T = \frac{E_T}{R_T}$$

$$I_T = \frac{120\ \text{V}}{222.2\ \Omega}$$

$$I_T = 0.54\ \text{A}$$

36. Determine the amount of current flow through each resistor.

$$I_1 = \frac{E_1}{R_1}$$

$$I_1 = \frac{120\ \text{V}}{400\ \Omega}$$

$$I_1 = 0.3\ \text{A}$$

$$I_C = \frac{E_C}{R_C}$$

$$I_C = \frac{120\ \text{V}}{500\ \Omega}$$

$$I_C = 0.24\ \text{A}$$

37. Determine the amount of voltage dropped across each resistor. Resistor R_1 is connected in parallel with the output of the power supply. Its voltage drop is, therefore, the output voltage of the power supply, or 120 V. Resistor R_C is actually the combination of resistors R_2 and R_3, so the Ohm's law values computed for resistor R_C apply to resistors R_2 and R_3. Resistors R_2 and R_3 are connected in series with each other. One of the rules for a series circuit states that

the current flow in a series circuit is the same at all points in the circuit. Therefore, the amount of current flowing through resistor R_C is the current flowing through resistors R_2 and R_3. The amount of voltage dropped by these two resistors can now be computed using Ohm's law.

$$E_2 = I_2 \times R_2$$
$$E_2 = 0.24\ A \times 200\ \Omega$$
$$E_2 = 48\ V$$
$$E_3 = I_3 \times R_3$$
$$E_3 = 0.24\ A \times 300\ \Omega$$
$$E_3 = 72\ V$$

NOTE: One of the rules for a series circuit states that the total voltage is equal to the sum of the voltage drops. If the values of E_2 and E_3 are added, they will equal the voltage applied to them, or 120 V.

Circuit summary: The values for the circuit shown in Figure 5-3 are as follows:

E_T = 120 VAC	E_1 = 120 VAC	E_2 = 48 VAC	E_3 = 72 VAC
I_T = 0.54 A	I_1 = 0.3 A	I_2 = 0.24 A	I_3 = 0.24 A
R_T = 222.2 Ω	R_1 = 400 Ω	R_2 = 200 Ω	R_3 = 300 Ω

38. Disconnect the circuit from terminals 4 and N on the power supply.

39. Measure the total resistance of the circuit with an ohmmeter.

    R_T = ______________ Ω

40. Compare this measured value with the computed value.

41. Reconnect the circuit to terminals 4 and N of the power supply.

42. Turn on the power and adjust the voltage until voltmeter V_T indicates a value of 120 V.

43. Measure the current flowing through ammeters A_T, A_1, and A_2.

    NOTE: Change the value of ammeter A_T to 2.5 A

    A_T______________ A A_1______________ A A_2______________ A

44. Compare these measured values with the computed values.

45. Measure the voltage drops across the resistors in the circuit using voltmeters V_T, V_2, and V_3.

    V_T______________ V V_2______________ V V_3______________ V

46. Compare these measured values with the computed values.

47. **Return the voltage to 0 V and turn off the power supply.**

48. Using the variable-resistance module, change the value of resistor R_1 to 300 Ω, resistor R_2 to 200 Ω, and resistor R_3 to 400 Ω.

49. Compute the combined resistance of resistors R_2 and R_3 by adding their values together.

$$R_C = R_2 + R_3$$

    R_C = ______________ Ω

50. Compute the total resistance of the circuit using the formula

$$R_T = \frac{R_1 \times R_C}{R_1 + R_C}$$

R_T = ______________ Ω

51. Assume that a voltage of 120 V is connected to the circuit and compute the total current flow in the circuit using Ohm's law.

I_T = ______________ A

52. Determine the amount of current flow through resistors R_1 and R_C using Ohm's law.

I_1 = ______________ A I_C = ______________ A

53. Because resistor R_C is the combination of resistors R_2 and R_3, the amount of current flowing through resistor R_C is the same as the current flowing through resistors R_2 and R_3. Find the amount of voltage dropped by resistors R_2 and R_3 using Ohm's law.

E_2 = ______________ V E_3 = ______________ V

54. Disconnect the circuit from terminals 4 and N of the power supply.

55. Measure the total resistance of the circuit using an ohmmeter.

R_T = ______________ Ω

56. Compare this measured value with the computed value.

57. Reconnect the circuit to terminals 4 and N of the power supply.

58. Turn on the power supply and adjust the voltage for a value of 120 V.

59. Measure the amount of current flowing through ammeters A_T, A_1, and A_2.

A_T______________ A A_1______________ A A_2______________ A

60. Compare these measured values with the computed values.

61. Measure the voltage drops indicated by voltmeters V_T, V_2, and V_3.

V_T______________ V V_2______________ V V_3______________ V

62. Compare these measured values with the computed values.

63. **Return the voltage to 0 V and turn off the power supply.**

64. Disconnect the circuit and return the components to their proper place.

Review Questions

1. State the three rules for series circuits.

A. ____

B. ____

C. ____

2. State the three rules for parallel circuits.

   A. ______________________________

   B. ______________________________

   C. ______________________________

3. Refer to Figure 5-1. Assume that resistor R_1 has a value of 1200 Ω, resistor R_2 has a value of 2400 Ω, and resistor R_3 has a value of 3000 Ω. What is the total resistance of the circuit?

   R_T = ____________ Ω

4. Assume that a voltage of 24 VDC is applied to the circuit in question 3. What is the total current flow in the circuit?

   I_T = ____________ A

5. Using the values given in questions 3 and 4, compute the voltage drop across resistor R_1, the amount of current flow through resistors R_2 and R_3, and the voltage drop across resistors R_2 and R_3.

   E_1 = ____________ V E_2 = ____________ V E_3 = ____________ V

   I_1 = ____________ A I_2 = ____________ A I_3 = ____________ A

6. Refer to Figure 5-3. Assume that the resistance value of resistor R_1 is 510 Ω, the value of resistor R_2 is 270 Ω, and that of resistor R_3 is 330 Ω. What would be the total resistance of the circuit?

   R_T = ____________ Ω

7. Assume that a voltage of 60 VDC is applied to the circuit in question 6. What would be the total current flow in the circuit?

   I_T = ____________ A

8. Using the values given in questions 6 and 7, compute the current flow through each resistor and the voltage drop across each resistor.

   E_1 = ____________ V E_2 = ____________ V E_3 = ____________ V

   I_1 = ____________ A I_2 = ____________ A I_3 = ____________ A

Exercise 6

Alternating Current Resistive Circuits

Objectives

After completing this lab you should be able to:

- Discuss the operating characteristics of a pure resistive circuit when connected to alternating current.
- Connect a resistive circuit and make electrical measurements with measuring instruments.

Materials and Equipment

Power supply module	EMS 8821
Resistance load module	EMS 8311
AC voltmeter module	EMS 8426
Single-phase wattmeter module	EMS 8431
AC ammeter module	EMS 8425

Discussion

When a pure resistive load is connected to an alternating current circuit, the voltage and current are in phase with each other. Because there is no phase angle difference between the voltage and current, the power used by the load can be computed by multiplying the voltage applied to the resistor by the amount of current flow through the resistor. The formulas for computing power in an AC circuit are:

$$P = E \times I \times \cos \angle \Theta$$

$$P = I^2 \times R \times \cos \angle \Theta$$

$$P = \frac{E^2}{R} \times \cos \angle \Theta$$

Notice that in each formula, the values of voltage, current, or resistance must be multiplied by the cosine of angle theta. Because voltage and current are in phase in a pure resistive circuit, the value of angle theta is zero. The cosine of zero degrees is 1. Therefore, if the formula $P = E \times I \times \cos \angle \Theta$ is used to calculate power, the product of the voltage and current will be multiplied by 1.

Exercise 9

Resistive-Inductive Parallel Circuits

Objectives

After completing this lab you should be able to:

- Discuss the operation of a parallel circuit containing resistance and inductance.
- Compute circuit values of an RL parallel circuit.
- Connect an RL parallel circuit and measure circuit values with test instruments.

Materials and Equipment

Power supply module	EMS 8821
Variable-resistance module	EMS 8311
Variable-inductance module	EMS 8321
AC ammeter module	EMS 8425
Single-phase wattmeter module	EMS 8431
AC voltmeter module	EMS 8426

Discussion

When resistance and inductance are connected in parallel, the voltage across each device will be in phase with and have the same value as the other device. The current flow through the inductor will be 90° out of phase with the current flow through the resistor, however. The amount of phase angle shift between the total circuit current and voltage is determined by the ratio of the amount of resistance to the amount of inductance. The circuit power factor is still determined by the ratio of resistance and inductance.

The example circuit shown in Figure 9-1 has a resistance of 15 Ω and an inductive reactance of 20 Ω. The circuit is connected to a voltage of 240 VAC and a frequency of 60 Hz. In this example problem, the following circuit values will be computed:

I_R—current flow through the resistor
P— watts (true power)
I_L—current flow through the inductor
VARs—reactive power
I_T—total circuit current
Z—total impedance of the circuit
VA—volt-amperes (apparent power)
PF—power factor
$\angle\theta$ —angle theta

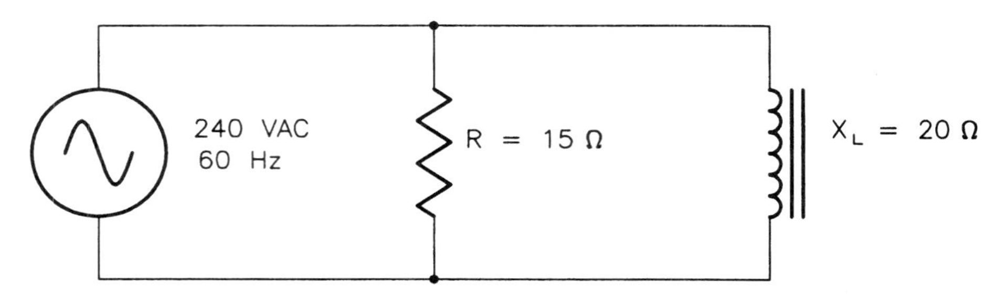

FIGURE 9-1 Resistive-inductive parallel circuit

The amount of current flow through the resistor will be computed by using the formula

$$I_R = \frac{E}{R}$$

$$I_R = \frac{240\ \text{V}}{15\ \Omega}$$

$$I_R = 16\ \text{A}$$

The amount of total power in the circuit will be computed using the formula

$$P = E \times I_R$$

$$P = 240\ \text{V} \times 16\ \text{A}$$

$$P = 3840\ \text{W}$$

The amount of current flow through the inductor will be computed using the formula

$$I_L = \frac{E}{X_L}$$

$$I_L = \frac{240\ \text{V}}{20\ \Omega}$$

$$I_L = 12\ \text{A}$$

The amount of reactive power, VARs, will be computed using the formula

$$\text{VARs} = E \times I_L$$

$$\text{VARs} = 240\ \text{V} \times 12\ \text{A}$$

$$\text{VARs} = 2880$$

The total current flow through the circuit can be computed by adding the current flow through the resistor and the current flow through the inductor together. Since these two currents are 90° out of phase with each other, vector addition will be used.

$$I_T = \sqrt{I_R^2 + I_L^2}$$
$$I_T = \sqrt{16^2 + 12^2}$$
$$I_T = \sqrt{256 + 144}$$
$$I_T = \sqrt{400}$$
$$I_T = 20 \text{ A}$$

The total impedance of the circuit will be computed using the formula

$$Z = \frac{E}{I_T}$$
$$Z = \frac{240 \text{ V}}{20 \text{ A}}$$
$$Z = 12\ \Omega$$

The apparent power is computed by multiplying the circuit voltage by the total current flow.

$$VA = E \times I_T$$
$$VA = 240 \text{ V} \times 20 \text{ A}$$
$$VA = 4800$$

The circuit power factor can be computed using the formula

$$PF = \frac{W}{VA} \times 100$$
$$PF = \frac{3840 \text{ W}}{4800 \text{ VA}} \times 100$$
$$PF = 0.80 = 80\%$$

Angle theta is the arc-cosine of the power factor.

$$\cos \angle \theta \quad 0.80$$
$$\angle \theta \quad 36.87$$

Name ______________________________ Date ______________

Procedure

1. Connect the circuit shown in Figure 9-2. If necessary, connect the wattmeter module to the 24-volt supply located on the power supply module. (Some wattmeter modules do not require this connection.)
2. Adjust the variable-resistance module for a resistance of 75 Ω.
3. Adjust the variable-inductance module for an inductive reactance of 80 Ω.
4. Turn on the power supply and adjust the voltage control rheostat until a voltage of 120 V is applied to the circuit.
5. Compute the current flow through the 75-Ω resistor using the formula

$$I_R = \frac{E}{R}$$

I_R = ______________ A

6. Measure the amount of current flow through the 75-Ω resistance and compare this value with the computed value.

I_R = ______________ A

7. Compute the amount of true power in this circuit using the formula

$$P = E \times I_R$$

P = ______________ W

8. Measure the amount of power in the circuit and compare the measured value with the computed value.

P = ______________ W

9. Compute the amount of current flow through the inductor using the formula

$$I_L = \frac{E}{X_L}$$

I_L = ______________ A

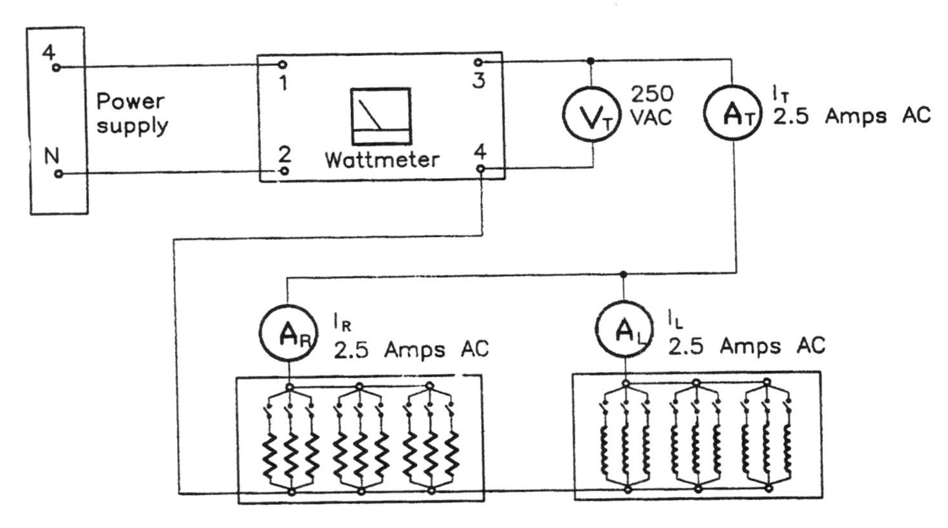

FIGURE 9-2 Connecting an RL parallel circuit

10. Measure the amount of current flow through the inductor and compare this value with the computed value.

    I_L = ____________ A

11. Compute the amount of total current flow in this circuit using the formula

$$I_T = \sqrt{I_R^2 + I_L^2}$$

    I_T = ____________ A

12. Measure the amount of total current flow in the circuit and compare this value with the computed value.

    I_T = ____________ A

13. Compute the amount of apparent power in the circuit using the formula

$$VA = E \times I_T$$

    VA = ____________

14. Compute the power factor for this circuit using the formula

$$PF = \frac{W}{VA} \times 100$$

    PF = ____________ %

15. Compute angle theta using the formula

$$\cos \angle \theta = PF$$

    $\angle\theta$ = ____________ °

16. **Return the voltage to 0 V and turn off the power supply.**
17. Disconnect the circuit and return the components to their proper place.

Review Questions

1. When an inductor and resistor are connected in parallel, how many degrees out of phase is the current flow through the resistor with the current flow through the inductor?

   ____________ °

2. An inductor and resistor are connected in parallel to a 120-V 60-Hz line. The resistor has a resistance of 50 Ω, and the inductor has an inductance of 0.2 H. What is the total current flow through the circuit?

   I_T = ____________ A

3. What is the impedance of the circuit in question 2?

   Z = ____________ Ω

4. What is the power factor of the circuit in question 2?

   PF = ____________ %

5. How many degrees out of phase are the current and voltage in question 2?

   $\angle\theta$ = ____________ °

Exercise 15

Three-Phase Circuits

Objectives

After completing this lab you should be able to:

- Discuss the differences between three-phase and single-phase voltages.
- Discuss the characteristics of delta and wye connections.
- Compute voltage and current values for delta and wye circuits.
- Connect delta and wye circuits and make measurements with measuring instruments.

Materials and Equipment

Power supply module	EMS 8821
Variable-resistance module	EMS 8311
AC ammeter module	EMS 8425
AC voltmeter module	EMS 8426

Discussion

A single-phase alternating voltage can be produced by rotating a magnetic field through the conductors of a stationary coil as shown in Figure 15-1.

Because alternate polarities of the magnetic field cut through the conductors of the stationary coil, the induced voltage will change polarity at the same speed as the rotation of the magnetic field. The alternator shown in Figure 15-1 is single-phase because it produces only one AC voltage.

If three separate coils are spaced 120° apart as shown in Figure 15-2, three voltages 120° out of phase with each other will be produced when the magnetic field cuts through the coils. This is the manner in which a three-phase voltage is produced. There are two basic three-phase connections: the wye, or star, and the delta. The wye, or star,

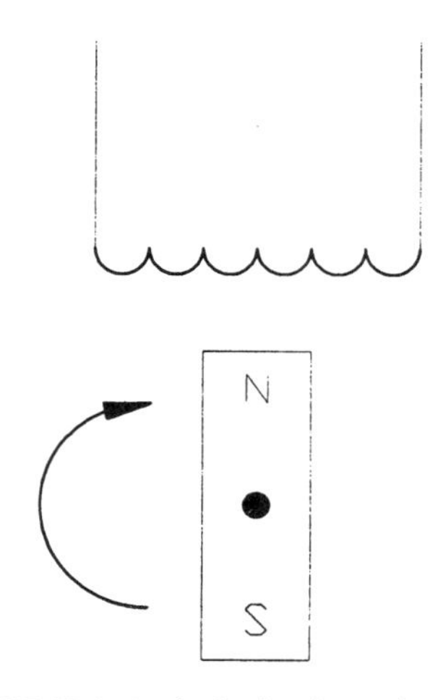

FIGURE 15-1 Basic single-phase alternator

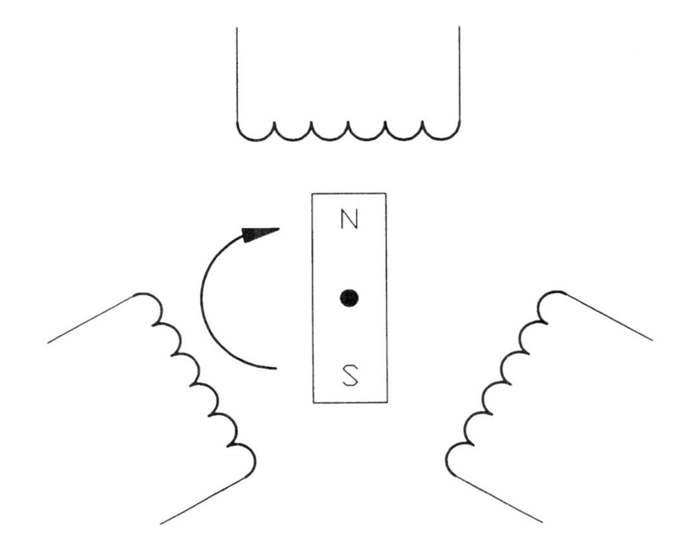

FIGURE 15-2 Basic three-phase alternator

connection is made by connecting one end of each of the three phase windings together as shown in Figure 15-3. The voltage produced by a single winding or phase is known as the phase voltage, as shown in Figure 15-4. The voltage measured between the lines is known as the line-to-line voltage, or simply as the line voltage.

In Figure 15-5, ammeters have been placed in the phase winding of a wye-connected load and in the line supplying power to the load. Voltmeters have been connected across the input to the load and across the phase. A line voltage of 208 V has been applied to the load. Notice that the voltmeter connected across the lines indicates a value of 208 V, but the voltmeter connected across the phase indicates a value of 120 V.

In a wye-connected system, the line voltage is higher than the phase voltage by a factor of the square root of 3 (1.732). Two formulas used to compute the voltage in a wye-connected system are:

$$E_{LINE} = E_{PHASE} \times 1.732$$

$$E_{PHASE} = \frac{E_{LINE}}{1.732}$$

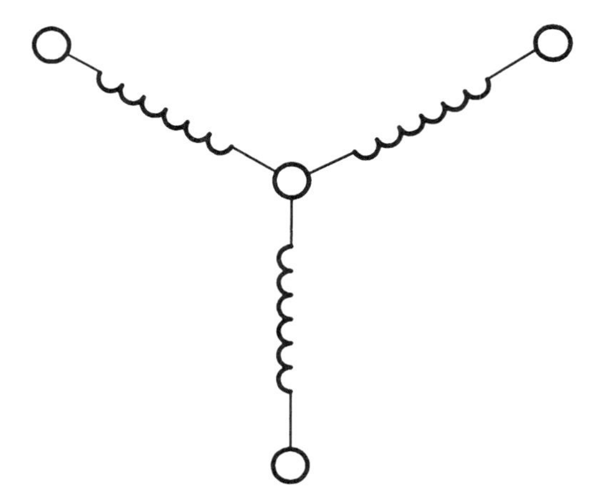

FIGURE 15-3 Wye connection

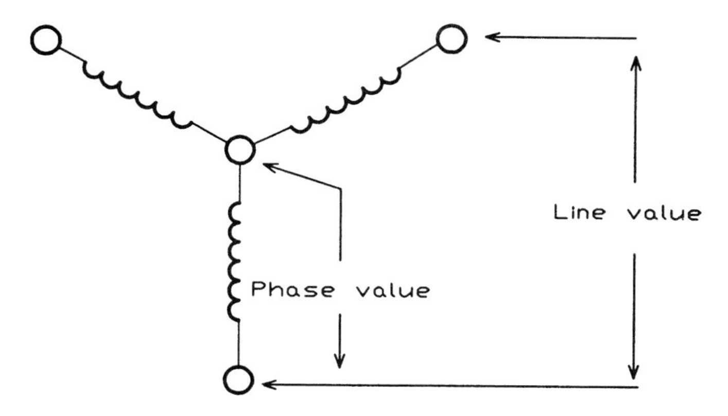

FIGURE 15-4 Line and phase voltage values of a wye connection

Notice in Figure 15-5 that there are 10 A of current flow in both the phase and the line. In a wye-connected system, phase current and line current are the same.

$$I_{LINE} = I_{PHASE}$$

In Figure 15-6, three separate loads have been connected to form a delta connection. This connection receives its name from the fact that in a schematic diagram it looks like the Greek letter delta (Δ). Voltmeters have been connected across the lines and across the phase. Ammeters have been connected in the line and in the phase. In the delta connection, line voltage and phase voltage are the same.

$$E_{LINE} = E_{PHASE}$$

Notice that the line current and phase current are different, however. The line current of a delta connection is higher than the phase current by a factor of the square root of 3 (1.732). Formulas for determining the current in a delta connection are:

$$I_{LINE} = I_{PHASE} \times 1.732$$

$$I_{PHASE} = \frac{I_{LINE}}{1.732}$$

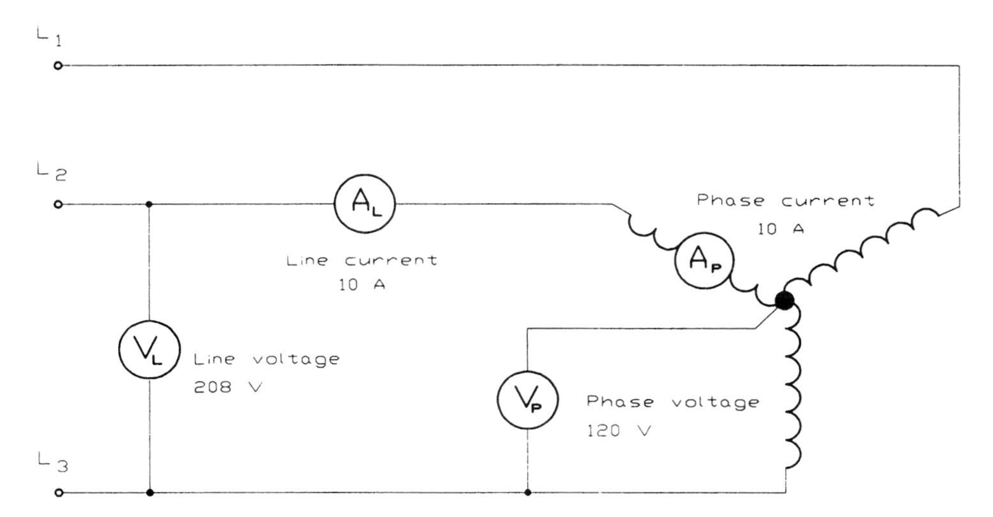

FIGURE 15-5 Wye-connected load

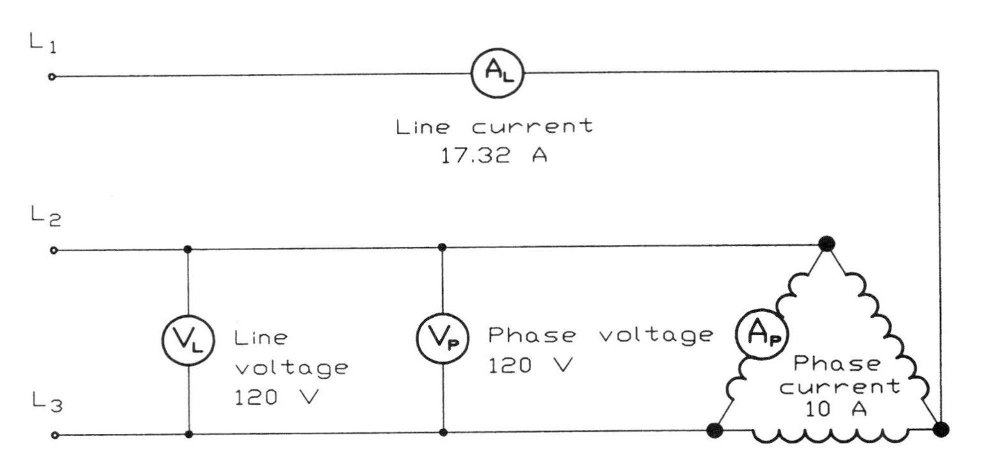

FIGURE 15-6 Delta-connected load

Name ______________________________ Date ______________

Procedure

1. Connect the circuit shown in Figure 15-7.
2. Set the variable-resistance module for a resistance of 200 Ω on each of the three phases. Notice that one end of each of the resistance banks have been connected together. This forms a wye connection.
3. Compute the voltage drop across each phase if a line voltage of 208 V is applied to the resistance bank.

$$E_{PHASE} = \frac{E_{LINE}}{1.732}$$

E_{PHASE} = ______________ V

4. Compute the amount of phase current that should flow in this circuit.

$$I_{PHASE} = \frac{E_{PHASE}}{R}$$

I_{PHASE} = ______________ A

5. Compute the amount of line current that should flow in this circuit.

$$I_{LINE} = I_{PHASE}$$

I_{LINE} = ______________ A

6. Turn on the power supply and adjust the output voltage for a line-to-line value of 208 V.
7. Measure the phase value of voltage and compare this with the computed value.

E_{PHASE} = ______________ V

8. Measure the line current and compare this with the computed value.

I_{LINE} = ______________ A

9. **Return the voltage to 0 V and turn off the power supply.**

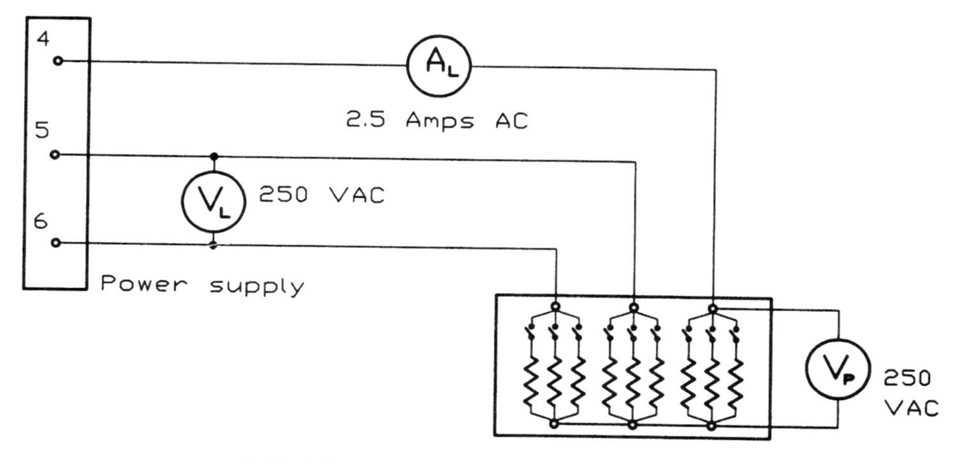

FIGURE 15-7 Wye-connected resistive load module

10. Reconnect the circuit to form a delta connection as shown in Figure 15-8.
11. A resistance value of 200 Ω per phase should be maintained.
12. Compute the value of phase voltage across the resistors if a line voltage of 120 V is connected to the circuit.

$$E_{PHASE} = E_{LINE}$$

E_{PHASE} = ______________ V

13. Compute the amount of current flow through the phase.

$$I_{PHASE} = \frac{E_{PHASE}}{R}$$

I_{PHASE} = ______________ A

14. Compute the amount of line current.

$$I_{LINE} = I_{PHASE} \times 1.732$$

I_{LINE} = ______________ A

15. Turn on the power supply and adjust the output voltage for a line-to-line value of **120 V.**
16. Measure the phase voltage across the load and compare this with the computed value.

E_{PHASE} = ______________ V

17. Measure the line current and compare this value with the computed value.

I_{LINE} = ______________ A

18. **Return the voltage to 0 V and turn off the power supply.**
19. Disconnect the circuit and return the components to their proper place.

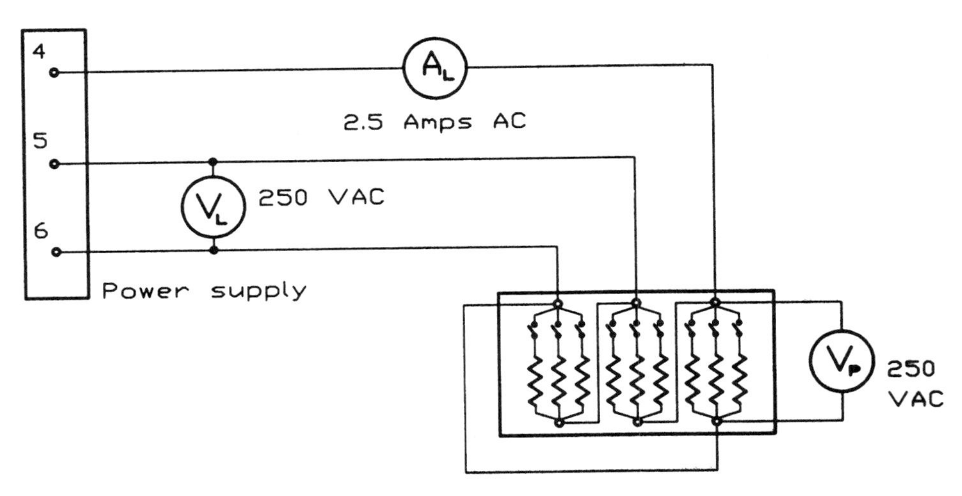

FIGURE 15-8 Delta-connected resistive load module

Review Questions

1. How many degrees out of phase with each other are the voltages of a three-phase system?

   ______________ °

2. What are the two main types of three-phase connections?

   ______________ and ______________

3. A wye-connected load has a voltage of 480 V applied to it. What is the voltage dropped across each phase?

   ______________ V

4. A wye-connected load has a phase current of 25 A. How much current is flowing through the lines supplying the load?

   ______________ A

5. A delta connection has a voltage of 560 V connected to it. How much voltage is dropped across each phase?

   ______________ V

6. A delta connection has 30 A of current flowing through each phase winding. How much current is flowing through each of the lines supplying power to the load?

   ______________ A

Exercise 16

Three-Phase Circuit Calculations

Objectives

After completing this lab you should be able to:

- Discuss power factor and phase angle calculations for three-phase circuits.
- Discuss the two-wattmeter method for measuring true power in a three-phase circuit.
- Calculate values of true power, apparent power, power factor, and phase angle for three-phase circuits.
- Make measurements of power and apparent power using measuring instruments.

Materials and Equipment

Power supply module	EMS 8821
Variable-resistance module	EMS 8311
Variable-capacitance module	EMS 8331
Variable-inductance module	EMS 8321
AC ammeter module	EMS 8425
Three-phase wattmeter module	EMS 8441
AC voltmeter module	EMS 8426

Discussion

The student of electric power technology often becomes confused when faced with computing values of voltage, current, impedance, true power, apparent power, reactive power, power factor, and phase angle for three-phase systems. One reason for this confusion is that more than one method can be used to compute three-phase values. The circuit shown in Figure 16-1 will be used to illustrate how true power or watts can be computed using two different methods.

In the circuit shown, three resistors having a value of 12 Ω each have been connected to form a wye connection. A line voltage of 208 V is being applied to the load. In the first method, the phase values of voltage and current will be used to compute true power, or watts. In a wye-connected system, the phase voltage is less than the line voltage by a factor of 1.732 (the square root of 3). The phase voltage, or the amount of voltage applied to each resistor, is:

$$E_{PHASE} = \frac{E_{LINE}}{1.732}$$

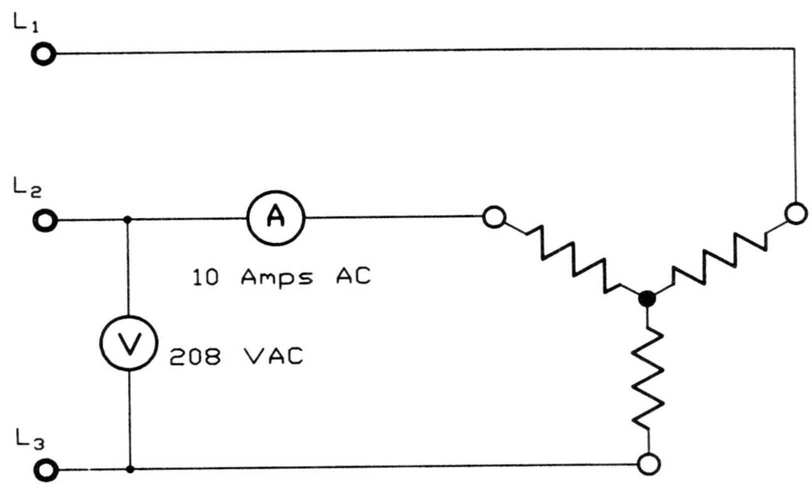

FIGURE 16-1 Three-phase wye connection

$$E_{PHASE} = \frac{208\ V}{1.732}$$
$$E_{PHASE} = 120\ V$$

Because each resistor has a voltage of 120 V applied to it, the current flow through each phase is:

$$I_{PHASE} = \frac{120\ V}{12\ \Omega}$$
$$I_{PHASE} = 10\ A$$

In a balanced three-phase system, the total power used is three times the amount of power used by one single phase. Therefore, when the phase values of voltage and current are used, power can be computed by using the formula

$$P = 3 \times E_{PHASE} \times I_{PHASE} \times PF$$
$$P = 3 \times 120\ V \times 10\ A \times 1$$
$$P = 3600\ W$$

NOTE: The power factor of a pure resistive circuit is unity, or 1.

The true power of a three-phase system can also be computed using the line values of voltage and current. In a wye-connected load, the phase current and line current are the same. The true power can be computed using the formula

$$P = \sqrt{3} \times E_{LINE} \times I_{LINE} \times PF$$
$$P = 1.732 \times 208\ V \times 10\ A \times 1$$
$$P = 3602.56\ W$$

The slight difference in answers between the two formulas is caused by rounding off values.

Notice that in the first example, the ***phase*** values of voltage and current are multiplied by 3. In the second example, the ***line*** values of voltage and current are multiplied by the square root of 3.

MEASURING THREE-PHASE POWER WITH WATTMETERS

When measuring the true power in a three-phase system, it is common to use two wattmeters connected as shown in Figure 16-2. When connecting wattmeters in this manner, care must be taken to notice the polarity connection of the voltmeter section as compared to the ammeter section of each wattmeter. In some instances, one of the wattmeters may indicate a reverse scale reading. When this condition occurs, the value of that wattmeter is considered to be a negative value. The power in the three-phase circuit is found by adding the values of the two wattmeters together. For example, assume that two wattmeters have been connected to a three-phase load, and one meter has an upscale indication of 125 W and the second meter has an upscale indication of 60 W. The total power in this circuit is:

$$W_T = W_1 + W_2$$
$$W_T = 125\ W + 60\ W$$
$$W_T = 185\ W$$

In the next example, assume that two wattmeters have been connected to a three-phase circuit. When the power is turned on, one of the wattmeters gives a reverse scale indication. In this case, the voltmeter connection for the reversed wattmeter would be changed to permit the meter to give an upscale reading. Now assume that the wattmeter that had to be reconnected indicates a value of 60 W and the other meter indicates a value of 125 W. In this example, the reversed meter reading must be considered a negative number.

$$W_T = W_1 + W_2$$
$$W_T = 125\ W + (-60\ W)$$
$$W_T = 125\ W - 60\ W$$
$$W_T = 65\ W$$

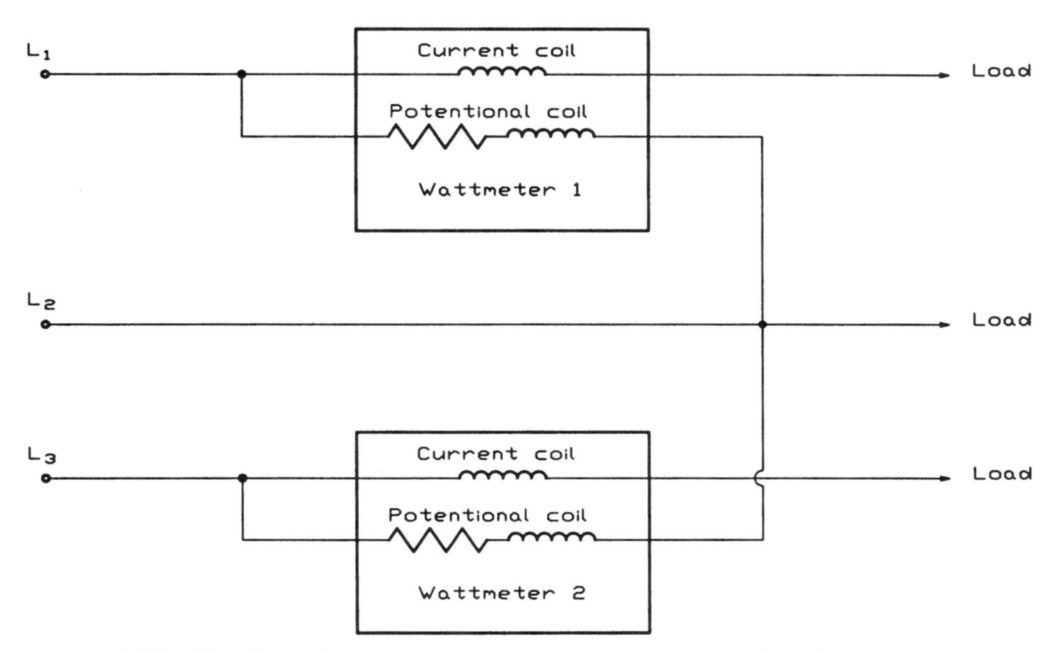

FIGURE 16-2 Connecting two single-phase wattmeters to measure three-phase power

THE EMS 8441 THREE-PHASE WATTMETER

The EMS 8441 three-phase wattmeter module, Figure 16-3, contains two wattmeters that have already been internally connected. Power is brought into the wattmeter module on terminals 1, 2, and 3. Power leaves the wattmeter module on terminals 4, 5, and 6. Each meter is connected to a switch labeled + and –. These switches are used to indicate when the meter is giving a forward reading or a reversed reading. When the wattmeter gives an upscale indication with the switch in the (+) position, the meter is indicating a forward reading. When the wattmeter gives an upscale reading with the switch in the (–) position, the meter is indicating a reverse reading. Therefore, when measuring three-phase power, if both wattmeters indicate upscale when both switches are in the (+) position, the two scale values are to be added together. If one switch is in the (+) position and the other switch is in the (–) position, the value of the wattmeter with the switch in the (–) position must be subtracted from the value of the wattmeter with the switch in the (+) position.

SIMPLIFYING THE CIRCUIT

When computing values in a three-phase system that contains both resistive and reactive components, it is often simpler to use the phase values and treat the circuit as a single-phase circuit. If phase values of voltage and current are used, the answers can be multiplied by 3 to find the three-phase value. For example, the circuit shown in Figure 16-4 contains resistive and inductive elements connected in parallel. In this circuit, three resistors having a value of 15 Ω each have been connected in a wye connection, and three inductors having an inductive reactance of 20 Ω each are connected in a wye connection. These two loads are connected in parallel to a 208-V line.

The amount of voltage connected across each phase can be computed using the formula

$$E_{PHASE} = \frac{E_{LINE}}{1.732}$$

$$E_{PHASE} = \frac{208\ V}{1.732}$$

$$E_{PHASE} = 120\ V$$

The circuit can now be treated as a single-phase circuit using phase values of voltage and current as shown in Figure 16-5. The amount of current flow through the resistor can be found using the formula

$$I_R = \frac{E_{PHASE}}{R}$$

$$I_R = \frac{120\ V}{15\ \Omega}$$

$$I_R = 8\ A$$

NOTE: I_R is the phase value of current, not the line value.

FIGURE 16-3 Three-phase wattmeter module. (Courtesy of Lab-Volt® Systems, Inc.)

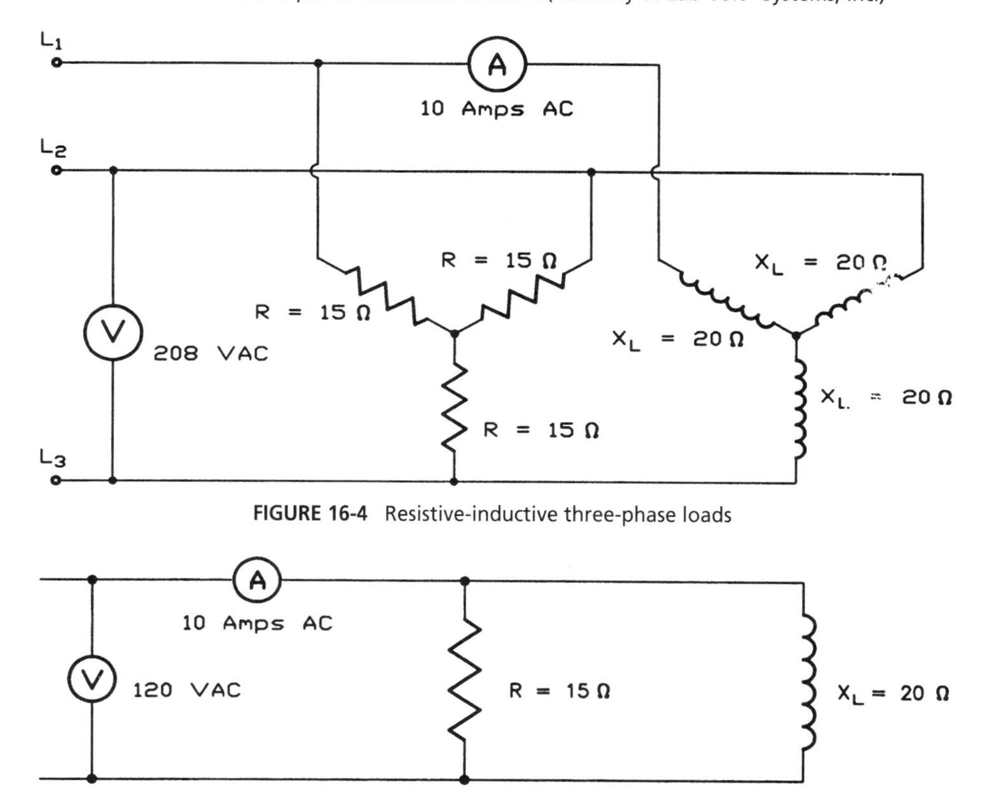

FIGURE 16-4 Resistive-inductive three-phase loads

FIGURE 16-5 Single-phase resistive-inductive parallel circuit

Now that the phase value of current and voltage for each phase of the resistive load is known, the three-phase value of true power can be computed using the formula

$$P = 3 \times E_{PHASE} \times I_R \times PF$$

$$P = 3 \times 120\ V \times 8\ A \times 1$$

$$P = 2880\ W$$

The amount of current flow through the inductors can be computed using the formula

$$I_L = \frac{E_{PHASE}}{X_L}$$
$$I_L = \frac{120\ V}{20\ \Omega}$$
$$I_L = 6\ A$$

NOTE: The letter *L* is used to indicate the value of ***inductive current,*** not the ***line*** value of current in a three-phase circuit. I_L is actually a ***phase*** value in this example. ***Line*** values of current will be shown as I_{LINE}.

$$VARs = 3 \times E_{PHASE} \times I_L$$
$$VARs = 3 \times 120\ V \times 6\ A$$
$$VARs = 2160$$

The amount of total current flow in each phase can be computed using the formula

$$I_T = \sqrt{I_R^2 + I_L^2}$$
$$I_T = \sqrt{8^2 + 6^2}$$
$$I_T = \sqrt{64 + 36}$$
$$I_T = \sqrt{100}$$
$$I_T = 10\ A$$

The amount of apparent power for the three-phase circuit can be computed using the formula

$$VA = 3 \times E_{PHASE} \times I_T$$
$$VA = 3 \times 120\ V \times 10\ A$$
$$VA = 3600$$

The power factor for this circuit can be computed using the formula

$$PF = \frac{W}{VA} \times 100$$
$$PF = \frac{2800\ W}{3600\ VA} \times 100$$
$$PF = 0.80 \times 100$$
$$PF = 80\%$$

Angle theta can be computed using the formula

$$\cos \angle\Theta = 0.80$$
$$\angle\Theta = 36.87^\circ$$

Name ______________________________ Date ______________

Procedure

1. Connect the circuit shown in Figure 16-6. If necessary, connect the wattmeter module to the 24-volt supply located on the power supply module. (Some wattmeter modules do not require this connection.)
2. Open (turn off) all switches on each of the three load modules.
3. Set each of the three banks of resistors on the variable-resistance module for a value of 200 Ω.

   NOTE: Each of the load modules contains three separate banks of devices. The resistive load module, for example, has a 1200-Ω, 600-Ω, and 300-Ω resistor connected to form each bank. Since this laboratory exercise deals with three-phase power, each of these separate banks will be used to form one phase of the three-phase load. To set the resistive load module for a value of 200 Ω per phase, close (turn on) the 600-Ω and 300-Ω switch on each of the three sections. The resistive load module now contains three 200-Ω resistors. The same procedure is to be followed when setting the inductive and capacitive load modules.

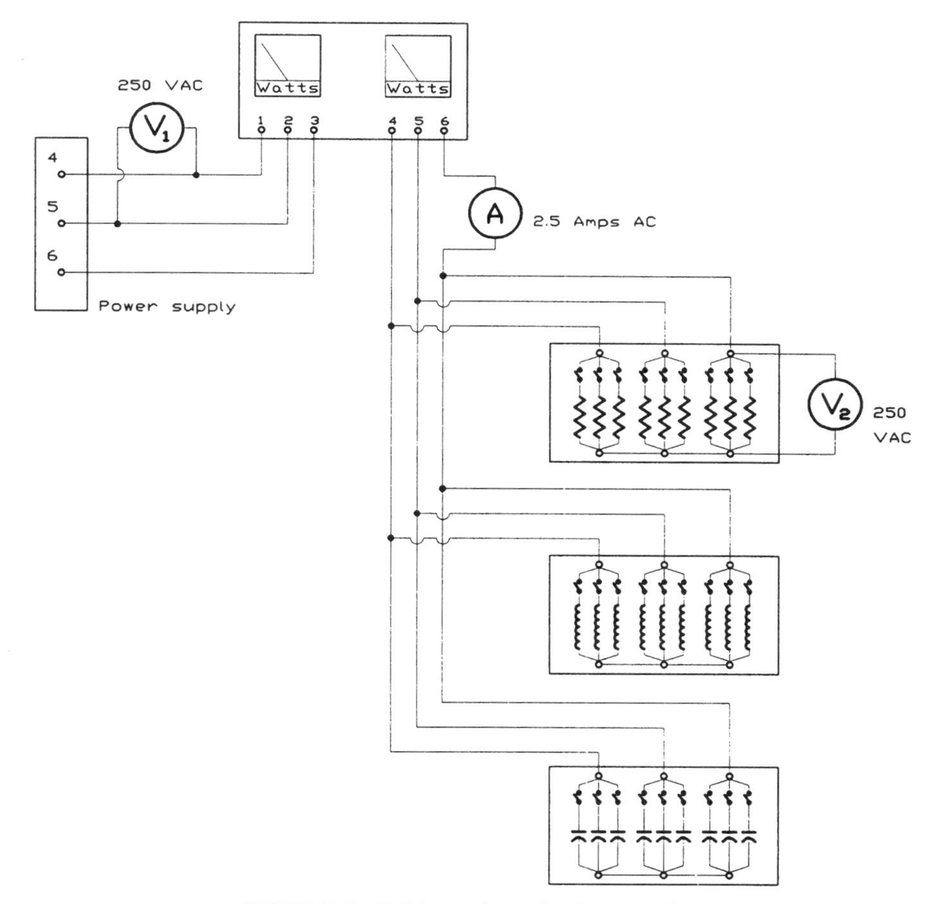

FIGURE 16-6 RLC three-phase circuit connection

4. Set each bank of inductors on the variable-inductance module for an inductive reactance of 300 Ω.

5. The voltage that is to be applied to the circuit will have a line value of 208 V. Compute the value of voltage that will be applied across each of the phases of the wye-connected loads.

$$E_{PHASE} = \frac{E_{LINE}}{1.732}$$

E_{PHASE} = ______________ V

6. Compute the amount of current flow through one phase of the resistor bank.

$$I_R = \frac{E_{PHASE}}{R}$$

I_R = ______________ A

7. Compute the amount of three-phase power in this circuit.

$$P = 3 \times E_{PHASE} \times I_R$$

P = ______________ W

8. Compute the amount of current flow through each phase of the inductor bank.

$$I_L = \frac{E_{PHASE}}{X_L}$$

I_L = ______________ A

9. Compute the amount of reactive power in the circuit.

$$VARs = 3 \times E_{PHASE} \times I_L$$

VARs = ______________

10. Compute the amount of total current flow in each phase using the formula

$$I_T = \sqrt{I_R^2 + I_L^2}$$

I_T ______________ A

11. Compute the apparent power in the circuit using the formula

$$VA = 3 \times E_{PHASE} \times I_T$$

VA = ______________

12. Compute the power factor for the circuit using the formula

$$PF = \frac{W}{VA} \times 100$$

PF = ______________ %

13. Compute angle theta using the formula

$$\cos \angle\Theta = PF$$

$\angle\Theta$ = ______________ °

14. Turn on the power supply and adjust the output voltage for a line value of 208 V.
15. Measure the voltage drop across the phase and compare this with the computed value.

    E_{PHASE} = ______________ V
16. Measure the true power in the circuit and compare this value with the computed value.

    P = ______________ W
17. Measure the total current flow in the circuit and compare this value with the computed value.

    I_T ______________ A
18. Use the measured values of line voltage and line current to compute the apparent power in the circuit and compare this value with the computed value in step 11.

$$VA = \sqrt{3} \times E_{LINE} \times I_{LINE}$$

    VA = ______________
19. **Return the voltage to 0 V and turn off the power supply.**
20. Reset the variable-resistance module so that each bank has a resistance of 600 Ω.
21. Open (turn off) all switches on the variable-inductance module.
22. Set the values of the variable-capacitance module so that each of the three banks has a capacitive reactance of 200 Ω.
23. Compute the amount of current flow through each phase of the resistive load. Assume a line voltage of 208 V.

    I_R = ______________ A
24. Compute the amount of power in the circuit using the phase values of voltage and current.

    P = ______________ W
25. Compute the amount of current flow through each of the phases in the capacitive load bank.

$$I_C = \frac{E_{PHASE}}{X_C}$$

    I_C = ______________ A
26. Compute the amount of reactive power in this circuit.

$$VARs = 3 \times E_{PHASE} \times I_C$$

    VARs = ______________

27. Compute the total amount of current flow in the circuit using the formula

$$I_T = \sqrt{I_R^2 + I_C^2}$$

I_T = _______________ A

28. Compute the apparent power in the circuit using the formula

$$VA = \sqrt{3} \times E_{LINE} \times I_{LINE}$$

NOTE: Because the loads are connected in a wye connection, I_T is both the ***line*** value and the ***phase*** value of current.

VA = _______________

29. Turn on the power supply and adjust the line voltage for a value of 208 V.

30. Measure the true power in the circuit and compare this value with the computed value.

$$W_T = W_1 + W_2$$

W_T = _______________ W

31. Measure the total current flow in the circuit and compare this value with the computed value.

I_T = _______________ A

32. Compute the value of apparent power using the measured values of line voltage and line current and compare this value with the computed value in step 28.

VA = _______________

33. **Return the voltage to 0 V and turn off the power supply.**

34. Leave the variable-resistance module set for a value of 600 Ω per phase. Reset the variable-inductance module for an inductive reactance value of 200 Ω per phase, and the variable-capacitance module for a capacitive reactance value of 400 Ω per phase.

35. Compute the amount of current flow through each phase of the resistance load. Assume a line voltage of 208 V.

I_R = _______________ A

36. Compute the amount of true power in this circuit.

P = _______________ W

37. Compute the amount of current flow through each phase of the inductive load.

I_L = _______________ A

38. Compute the amount of current flow through each phase of the capacitive load.

I_C = _______________ A

39. Compute the amount of total line current in each phase.

$$I_T = \sqrt{I_R^2 + (I_L - I_C)^2}$$

I_T = _______________ A

40. Turn on the power supply and adjust the line voltage for a value of 208 V.

41. Measure the amount of true power in the circuit and compare this value with the computed value.

    P_T = ______________ W

42. Measure the total current flow through each line and compare this with the computed value.

    I_T = ______________ A

43. Using the values of line voltage and line current compute the apparent power in this circuit.

    VA = ______________

44. Compute the power factor in this circuit.

    PF = ______________ %

45. Compute the total reactive power in this circuit using the formula

$$\text{VARs} = \sqrt{\text{VA}^2 - \text{W}^2}$$

    VARs = ______________

46. Compute angle theta for this circuit using the formula

$$\cos \angle\theta \; \text{PF}$$

    $\angle\theta$ = ______________ °

47. **Return the voltage to 0 V and turn off the power supply.**
48. Disconnect the circuit and return the components to their proper place.

Review Questions

1. A three-phase resistive load has a phase voltage of 440 V and a phase current of 25 A. What is the true power in this circuit?

   P = ______________ W

2. A three-phase resistive load is connected as a delta connection. The line voltage applied to the load is 560 V, and the current flow through the phase is 140 A. What is the true power in the circuit?

   P = ______________ W

3. A three-phase motor is connected to a 480-V line. A clamp-on ammeter indicates a current flow of 45 A in each of the three lines. What is the apparent power in this circuit?

   VA = ______________

4. Two wattmeters are connected to the motor in question 3. Both wattmeters show an upscale reading when connected to read in the forward direction. One wattmeter measures 15 kW and the other wattmeter measures 16.8 kW. What is the true power in the circuit?

   P = ______________ W

5. What is the power factor of the motor in question 3?

   PF = ______________ %

Exercise 17

Single-Phase Transformers: Part 1

Objectives

After completing this lab you should be able to:

- Discuss the different types of transformers.
- Calculate values of voltage, current, and turns for single-phase transformers.
- Test a transformer with an ohmmeter for opens or grounds.
- Connect a transformer and test the voltage output of different windings.

Materials and Equipment

Power supply module	EMS 8821
Transformer module	EMS 8341
AC voltmeter module	EMS 8426
Ohmmeter (supplied by student)	

Discussion

A transformer is a magnetically operated machine that can change values of voltage, current, and impedance without a change of frequency. Transformers are the most efficient machines known. Their efficiencies commonly range from 90% to 99%. Transformers can be divided into three classifications: isolation, auto, and current.

All values of a transformer are proportional to their turns ratio. This does not mean that the exact number of turns of wire on each winding must be known to determine different values of voltage and current for a transformer. What must be known is the ratio of turns. For example, assume a transformer has two windings. One winding, the primary, has 1000 turns of wire and the other winding, the secondary, has 250 turns of wire. The turns ratio of this transformer is 4 to 1, or 4:1 (1000/250 = 4). This indicates that there are four turns of wire on the primary for every one turn of wire on the secondary.

TRANSFORMER FORMULAS

Different formulas can be used to find the values of voltage and current for a transformer. The following is a list of standard formulas:

where

N_P — number of turns in the primary

N_S — number of turns in the secondary

E_P — voltage of the primary

E_S — voltage of the secondary

I_P — current in the primary

I_S — current in the secondary

FIGURE 17-1 Isolation transformer

$$\frac{E_P}{E_S} = \frac{N_P}{N_S} \qquad \frac{E_P}{E_S} = \frac{I_S}{I_P} \qquad \frac{N_P}{N_S} = \frac{I_S}{I_P}$$

$$E_P \times N_S = E_S \times N_P \qquad E_P = I_P \times E_S \qquad I_S \times N_P = I_P \times N_S \qquad I_S$$

The primary winding of a transformer is the power input winding. It is the winding that is connected to the incoming power supply. The secondary winding is the load winding or output winding. It is the side of the transformer that is connected to the driven load, as in Figure 17-1.

ISOLATION TRANSFORMERS

The transformer shown in Figure 17-1 is an isolation transformer. This means that the secondary winding is physically and electrically isolated from the primary winding. There is no electrical connection between the primary and secondary winding. This transformer is magnetically coupled, not electrically coupled. This "line isolation" is often a very desirable thing. Since there is no electrical connection between the load and power supply, the transformer becomes a filter between the two. The isolation transformer will greatly reduce any voltage spikes that originate on the supply side before they are transferred to the load side. Some transformers are built with a turns ratio of 1:1. A transformer of this type will have the same input and output voltage and is used for the purpose of isolation only.

The values of voltage and current of the isolation transformer are determined by the turns ratio. For example, assume the isolation transformer shown in Figure 17-1 has 240 turns of wire on the primary and 60 turns of wire on the secondary. This is a ratio of 4:1 (240/60 = 4). Now assume that 120 V is connected to the primary winding. What is the voltage of the secondary winding?

$$\frac{E_P}{E_S} = \frac{N_P}{N_S}$$

$$\frac{120\text{ V}}{E_S} = \frac{240}{60}$$

$$240\ E_S = 7200$$

$$E_S = 30\text{ V}$$

The transformer in this example is known as a step-down transformer because it has a lower secondary voltage than primary voltage.

Now assume that the load connected to the secondary winding has an impedance of 5 Ω. The next problem is to calculate the current flow in the secondary and primary windings. The current flow of the secondary can be computed using Ohm's law because the voltage and impedance are known.

$$I = \frac{E}{Z}$$

$$I = \frac{30\ V}{5\ \Omega}$$

$$I = 6\ A$$

Now that the amount of current flow in the secondary is known, the primary current can be computed using the formula

$$\frac{E_P}{E_S} = \frac{I_S}{I_P}$$

$$\frac{120\ V}{30\ V} = \frac{6\ A}{I_P}$$

$$120 \times I_P = 180$$

$$I_P = 1.5\ A$$

Notice that the primary voltage is higher than the secondary voltage, but the primary current is much less than the secondary current.

NOTE: A good rule for transformers is that power in must equal power out.

The product of the primary voltage and current should equal the product of the secondary voltage and current.

Primary	Secondary
120 V × 1.5 A = 180 VA	30 V × 6 A = 180 VA

In the next example, assume that the primary winding contains 240 turns of wire and the secondary contains 1200 turns of wire. This is a turns ratio of 1:5 (1200/240 = 5). Now assume that 120 V is connected to the primary winding. Compute the voltage output of the secondary winding.

$$\frac{E_P}{E_S} = \frac{N_P}{N_S}$$

$$\frac{120\ V}{E_S} = \frac{240}{60}$$

$$240 \times E_S = 144,000$$

$$E_S = 600\ V$$

Notice that the secondary voltage of this transformer is higher than the primary voltage. This type of transformer is known as a step-up transformer.

Now assume that the load connected to the secondary has an impedance of 2400 Ω. Find the amount of current flow in the primary and secondary windings. The current flow in the secondary winding can be computed using Ohm's law.

$$I = \frac{E}{Z}$$

$$I = \frac{600\ V}{2400\ \Omega}$$

$$I = 0.25\ A$$

Now that the amount of current flow in the secondary is known, the primary current can be computed using the formula

$$\frac{E_P}{E_S} = \frac{I_S}{I_P}$$

$$\frac{120\ V}{600\ V} = \frac{0.25\ A}{I_P}$$

$$120 \times I_P = 150\ A$$

$$I_P = 1.25\ A$$

Notice that the amount of power input equals the amount of power output.

Primary	Secondary
120 V × 1.25 A = 150 VA	600 V × 0.25 A = 150 VA

AUTOTRANSFORMERS

Autotransformers are one-winding transformers. They use the same winding for both the primary and secondary, as shown in Figure 17-2. The primary winding is between points B and N and has a voltage of 120 V applied to it. If the turns of wire are counted between points B and N, it can be seen there are 120 turns of wire. Now assume that the selector switch is set to point D. The load is now connected between points D and N. The secondary of this transformer contains 40 turns of wire. The following formula can be used to compute the amount of voltage applied to the load.

$$\frac{E_P}{E_S} = \frac{N_P}{N_S}$$

$$\frac{120\text{ V}}{E_S} = \frac{120}{40}$$

$$120 \times E_S = 4800$$

$$E_S = 40\text{ V}$$

Assume that the load connected to the secondary has an impedance of 10 Ω. The amount of current flow in the secondary circuit can be computed using the formula

$$I = \frac{E}{Z}$$

$$I = \frac{40\text{ V}}{10\ \Omega}$$

$$I = 4\text{ A}$$

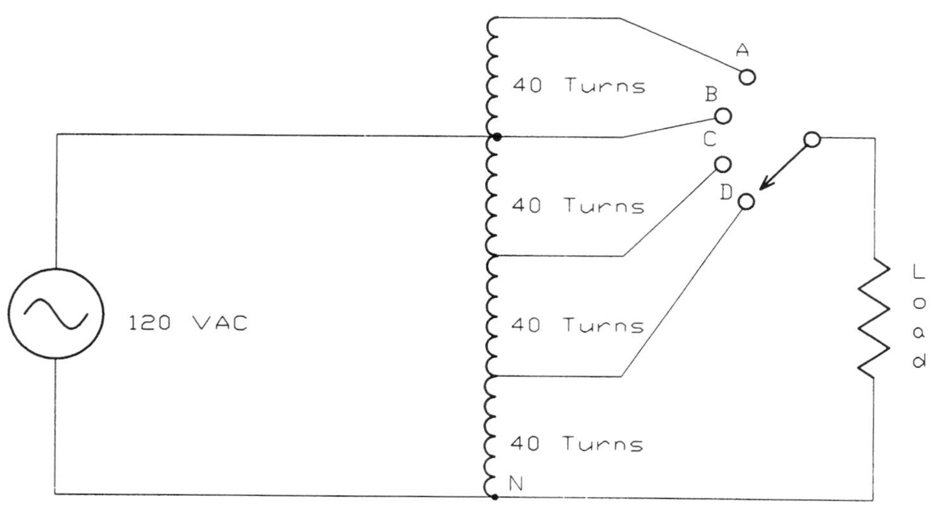

FIGURE 17-2 Autotransformer

The primary current can be computed by using the same formula that was used to compute primary current for an isolation transformer.

$$\frac{E_P}{E_S} = \frac{I_S}{I_P}$$
$$\frac{120\ V}{40\ V} = \frac{4\ A}{I_P}$$
$$120 \times I_P = 160\ A$$
$$I_P = 1.333\ A$$

The amount of power input and output for the autotransformer must be the same also.

Primary	Secondary
120 V × 1.333 A = 160	VA 40 V × 4 A = 160 VA

Now assume that the rotary switch is connected to point A. The load is now connected to 160 turns of wire. The voltage applied to the load can be computed by:

$$\frac{E_P}{E_S} = \frac{N_P}{N_S}$$
$$\frac{120\ V}{E_S} = \frac{120}{160}$$
$$120 \times E_S = 19{,}200$$
$$ES = 160\ V$$

Notice that the autotransformer, like the isolation transformer can be either a step-up or step-down transformer.

If the rotary switch shown in Figure 17-2 were to be removed and replaced with a sliding tap that made contact directly to the transformer winding, the turns ratio could be adjusted continuously. This type of transformer is commonly referred to as a "Variac," or "Powerstat." The Lab Volt power supply contains three of these autotransformers that are used to produce the variable voltage outputs.

The autotransformer does have one disadvantage. Because the load is connected to one side of the power line, there is no line isolation between the incoming power and the load. This can cause problems with certain types of equipment and must be a consideration when designing a power system.

CURRENT TRANSFORMERS

The current transformer is different from the voltage transformer in the manner of its connection. The primary winding of a voltage transformer is connected directly across the power line as shown in Figures 17-1 and 17-2. The primary winding of a current transformer is connected in series with the load as shown in Figure 17-3.

Current transformers are often used as instrumentation transformers in high-power circuits. Imagine the problem of measuring the amount of current flow in a power line designed to supply 20,000 A. It would be almost impossible to construct an ammeter that could connect in series with this type of circuit. The current transformer is used to help overcome this problem. Current transformers used for this application are designed to have a standard secondary current of 5 A when rated current flows through the primary, as shown in Figure 17-4. The current ratio of this transformer is, therefore,

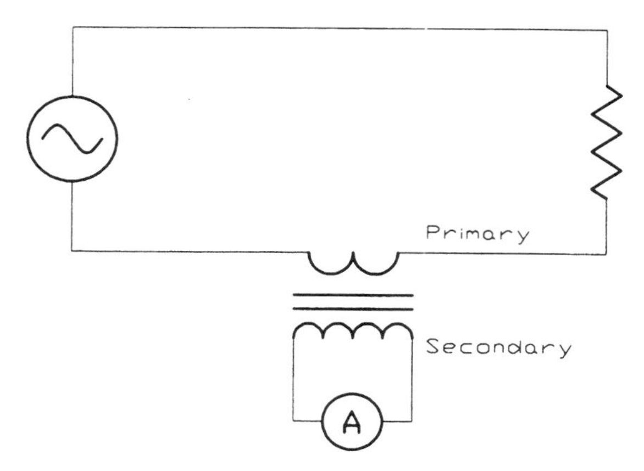

FIGURE 17-3 Current transformer

20,000:5. If the main power line were used as the primary winding, the number of primary turns would be 1. The number of turns of the secondary would be:

$$\frac{N_P}{N_S} = \frac{I_S}{I_P}$$

$$\frac{1}{N_S} = \frac{5\ \text{A}}{20,000\ \text{A}}$$

$$5 \times N_S = 20,000$$

$$N_S = 4000 \text{ turns}$$

The turns ratio of this transformer is 1:4000.

Notice that the secondary winding of this transformer is connected directly to an AC ammeter. Current transformers of this type are designed to have their secondary windings shorted. If the ammeter should be removed from the circuit while the main power is turned on, the current transformer would become a step-up voltage transformer with a turns ratio of 1:4000. For this reason, the following caution should always be kept in mind.

Caution: ***The secondary winding of a current transformer must never be opened when the main power is turned on. The secondary winding must always be shorted, or an extremely high voltage can be present at the secondary terminals. It is possible for this voltage to kill anyone who comes in contact with it.***

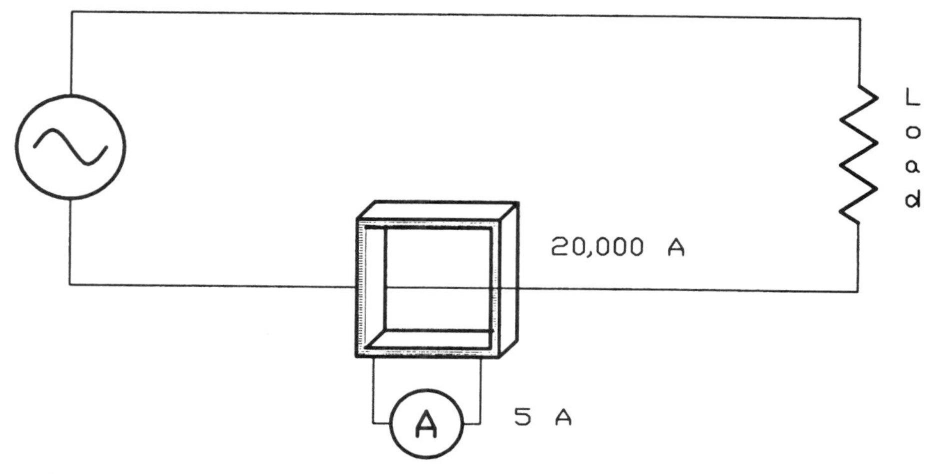

FIGURE 17-4 Current transformers used for instrumentation purposes have a standard secondary current of 5 A.

Name ______________________________ Date ______________

Procedure

1. Remove the EMS 8341 transformer module, Figure 17-5, from the cabinet.
2. Observe the connections made from the transformer to the banana plug jacks located on the front of the module.
3. Replace the transformer module in the cabinet. List the rated voltage and current between the points specified. The voltage and current ratings of each winding or section of winding is shown on the front of the transformer module.

   1 to 2__________ V __________ A

   3 to 7__________ V __________ A

   3 to 8__________ V __________ A

   3 to 4__________ V __________ A

   7 to 8__________ V __________ A

   7 to 4__________ V __________ A

   8 to 4__________ V __________ A

   5 to 9__________ V __________ A

   5 to 6__________ V __________ A

   9 to 6__________ V __________ A

4. Using an ohmmeter, test for continuity between the following terminals. Indicate whether there is continuity or no continuity. Write “yes” to indicate continuity and “no” to indicate no continuity.

   1 to 2__________

   1 to 3__________

   2 to 3__________

   3 to 7__________

   3 to 8__________

   3 to 4__________

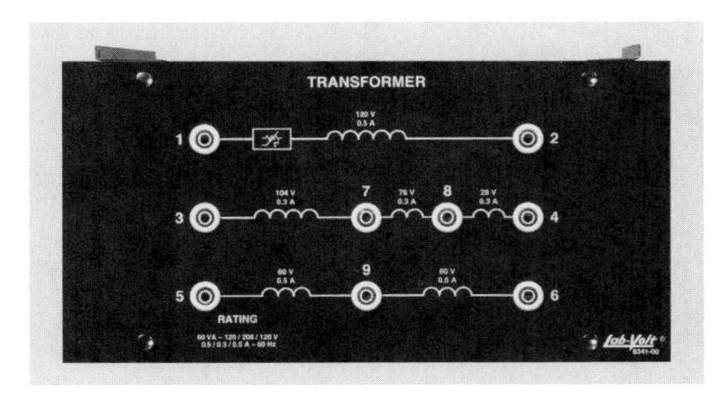

FIGURE 17-5 Single-phase Transformer Module. (Courtesy of Lab-Volt® Systems, Inc.)

3 to 5____________

3 to 6____________

7 to 8____________

7 to 4____________

8 to 4____________

4 to 5____________

4 to 6____________

5 to 9____________

5 to 6____________

9 to 6____________

5. Notice that the transformer contains three separate windings. One winding is between terminals 1 and 2, the second winding is between terminals 3 and 4, and the third winding is between terminals 5 and 6. The winding between points 3 and 4 has been tapped at two different points. These taps have been connected to terminals 7 and 8. The winding between points 5 and 6 has been "center-tapped." The center tap is connected to terminal 9. Is this transformer an isolation transformer or an autotransformer?

____

6. Connect the circuit shown in Figure 17-6.
7. Using a 250-VAC meter, measure the voltages between the listed terminals. The power should be turned off each time the voltmeter is reconnected. For voltages less than 100 V, the 100-V range of the meter may be used for greater accuracy. Compare these measured values with the values listed earlier.

1 to 2____________ V

3 to 7____________ V

3 to 8____________ V

3 to 4____________ V

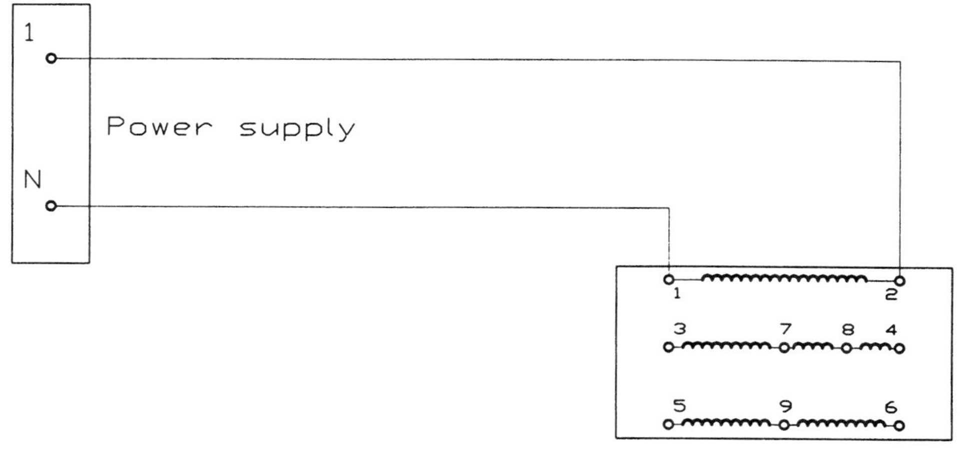

FIGURE 17-6 Connecting the transformer module

7 to 8______________ V

7 to 4______________ V

8 to 4______________ V

5 to 9______________ V

5 to 6______________ V

9 to 6______________ V

8. **Turn off the power supply and disconnect the circuit.**
9. Return the components to their proper place.

Review Questions

1. What is a transformer?

____

2. What are common efficiencies for transformers?

____

3. What is an isolation transformer?

____

4. All values of a transformer are proportional to its:

____

5. What is an autotransformer?

____

6. What is a disadvantage of an autotransformer?

____

7. What is the difference in connection between a voltage transformer and a current transformer?

____

8. What is the standard AC current rating of the secondary of a current transformer used for instrumentation?

______________ A

9. Why should the secondary of a current transformer used for instrumentation never be opened when the primary is energized?

____

10. Explain the difference between a step-up and a step-down transformer.

____

____

11. A transformer has a primary voltage of 240 V and a secondary voltage of 48 V. What is the turns ratio of this transformer?

____

Exercise 18

Single-Phase Transformers: Part 2

Objectives

After completing this lab you should be able to:

- Compute the turns ratio of a transformer.
- Compute voltage and current relationships for a transformer.
- Make measurements of voltage and current of the primary and secondary windings.

Materials and Equipment

Power supply module	EMS 8821
AC ammeter module	EMS 8425
AC voltmeter module	EMS 8426
Transformer module	EMS 8341
Resistance load module	EMS 8311

Discussion

All values of a transformer are proportional to its turns ratio. For example, assume that a transformer has a primary voltage of 240 V and a secondary voltage of 48 V. The turns ratio of this transformer can be found by dividing the highest voltage by the lowest voltage (240 V/48 V = 5). This transformer has a turns ratio of 5:1. There are 5 turns of wire in the primary winding for every 1 turn of wire in the secondary winding. Now assume that a load resistance of 4.8 Ω is connected to the secondary as shown in Figure 18-1. This will cause a current of 10 A to flow when the transformer is connected to the power supply (48 V/4.8 Ω = 10 A). Now determine the amount of current that will flow through the primary winding when the power is connected to the circuit.

$$\frac{E_P}{E_S} = \frac{I_S}{I_P}$$

$$\frac{240\text{ V}}{48\text{ V}} = \frac{10\text{ A}}{I_P}$$

$$240 \times I_P = 480\text{ A}$$

$$I_P = 2\text{ A}$$

Notice that only 2 A of current flow in the primary winding when there is 10 A of current in the secondary winding. Recall that the power input to a transformer must be equal to the power output. Therefore, volts times amps input must equal volts times amps output (240 V × 2 A = 480 VA) (48 V × 10 A = 480 VA).

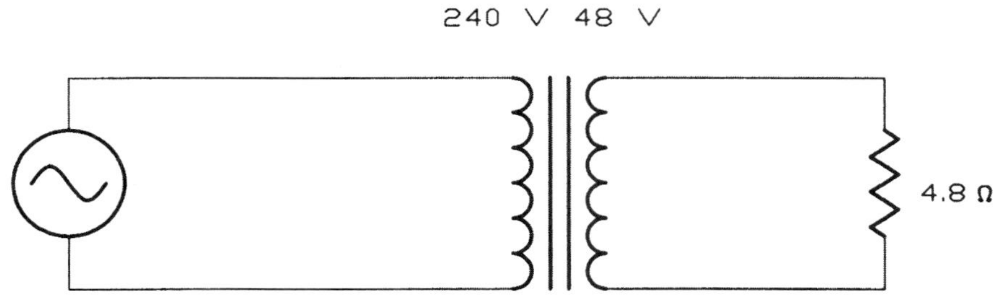

FIGURE 18-1 Step-down transformer

EXCITATION CURRENT

In actual practice, an ammeter would indicate slightly more than 2 A of current in the primary winding of the above-mentioned transformer. The reason for this is excitation current. The excitation current is the amount of current necessary to magnetize the iron core of the transformer. If there is no load connected to the secondary of the transformer, there will still be some excitation current supplied to the primary. The excitation current must be added to the current supplied to the secondary. For example, assume that the above transformer has an excitation current of 0.1 A. An ammeter connected to the primary winding would reveal that the actual current flow is 2.1 A. It should be noted, however, that the amount of excitation current is ordinarily so small compared with the full load current it is often omitted when making transformer calculations.

TRANSFORMER VOLTAGE DROP

It is not unusual for the secondary voltage of a transformer to drop below its rated value as it approaches full load. The reason for this voltage drop is the internal impedance of the transformer itself. The amount of internal impedance is determined by three factors:

1. the amount of iron used in the core of the transformer,
2. the size of the wire used to construct the windings, and
3. the number of turns of wire in the different windings.

Small transformers are often designed to produce higher-than-rated voltages at no load or light load. When load is added, the voltage will drop. It is intended that the voltage will drop to the rated secondary voltage at full load. Assume for example that a transformer has a primary voltage of 120 V and secondary rated at 12 V and 3 A. If the transformer is operated at no load, the secondary voltage may actually be 16 to 18 V. When the secondary is supplying the rated current of 3 A, however, the voltage should drop to 12 V.

In the following experiment, the transformer module will be connected to a load. The secondary voltage will drop below its rated value because of the internal impedance of the transformer.

Name ______________________________________ Date ______________

Procedure

1. Connect the circuit shown in Figure 18-2.
2. Compute the turns ratio of the transformer by dividing the primary voltage by the secondary voltage. (Voltages are shown on the module.)

   Ratio = ______________

3. Assume that a resistance of 240 Ω is to be connected to the secondary winding. Compute the amount of current flow.

$$I_S = \frac{E_S}{R}$$

   I_S = ______________ A

4. Now compute the amount of current that should flow in the primary winding.

$$\frac{E_P}{E_S} = \frac{I_S}{I_P}$$

   I_P = ______________ A

5. Set the resistance load module for a value of 240 Ω.
6. Turn on the power supply and adjust the output voltage for a value of 120 V.

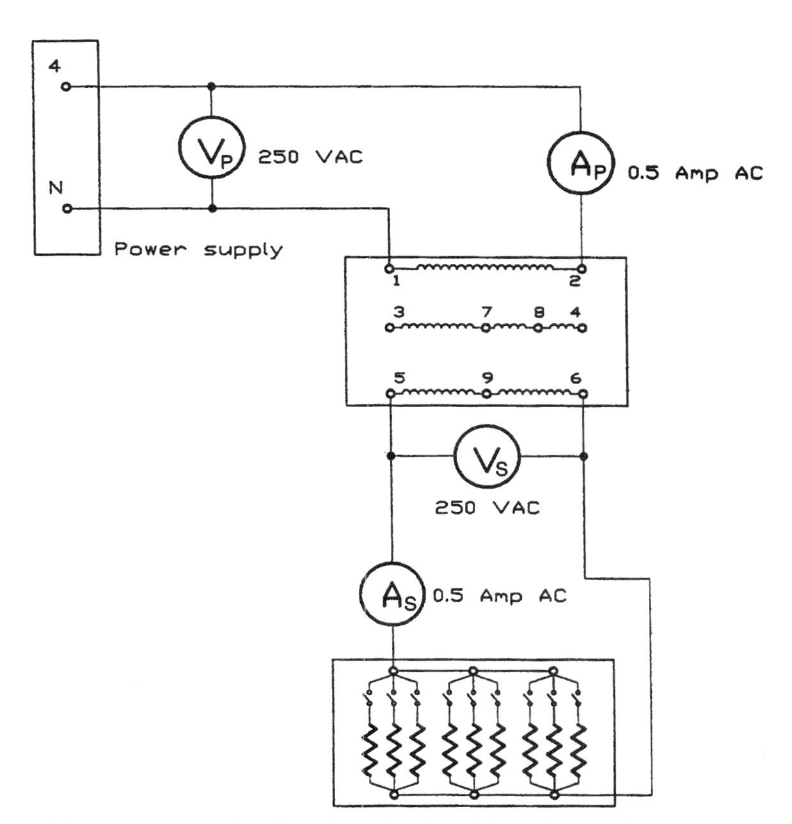

FIGURE 18-2 Connecting the resistive load module to the transformer module

7. Measure and record the following values.

   E_P (primary voltage) = _______________ V

   E_S (secondary voltage) = _______________ V

   I_P (primary current) = _______________ A

   I_S (secondary current) = _______________ A

8. Compare the measured values with the computed values.
9. **Return the voltage to 0 V and turn off the power supply.**
10. Change the secondary leads of the transformer from 5 and 6 to 5 and 9. Also reset the AC voltmeter connected to the secondary winding to the 100-V range, as in Figure 18-3.
11. Compute the turns ratio of the transformer by dividing the primary voltage by the secondary voltage.

    Ratio = _______________

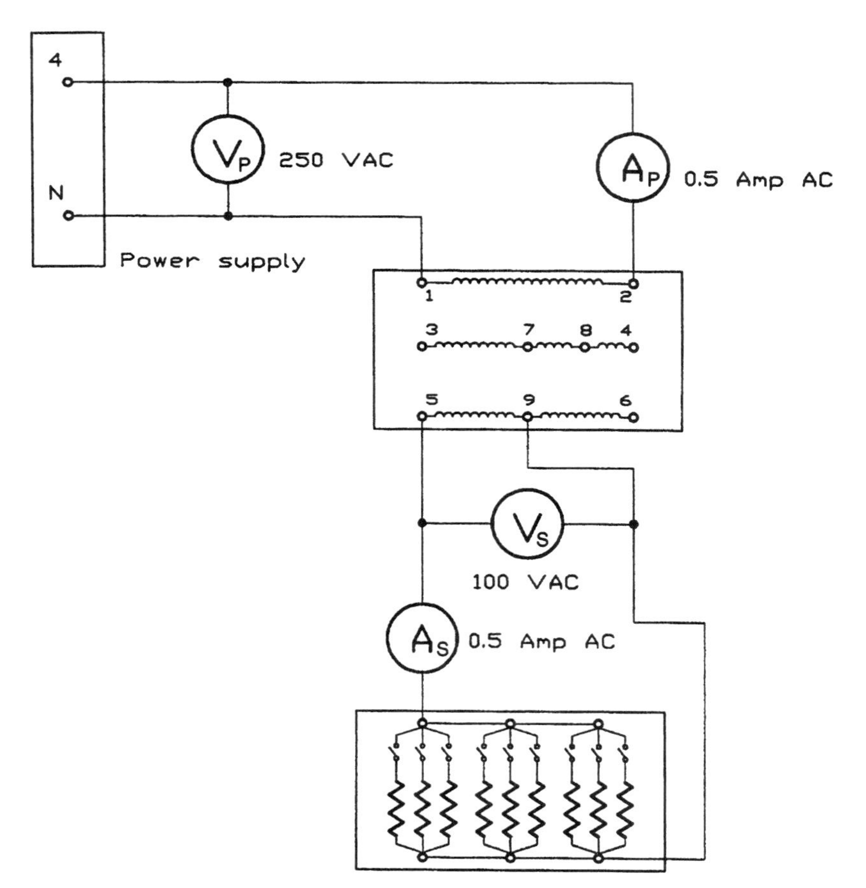

FIGURE 18-3 The secondary winding has been changed to 5 and 9.

12. Assume that a load of 120 Ω is to be connected to the secondary winding. Compute the amount of current flow through this winding.

$$I_S = \frac{E_S}{R}$$

I_S = ______________ A

13. Compute the amount of current flow through the primary winding using the formula

$$\frac{E_P}{E_S} = \frac{I_S}{I_P}$$

I_P = ______________ A

14. The same value for primary current can be found by using the turns ratio. In step 11 it was found that this transformer has a ratio of 2:1. Because the primary of the transformer has a higher voltage than the secondary, it will have a lower current. The amount of primary current will be proportional to the turns ratio (0.5/2 = 0.25 A). Notice that the answer is the same as that ascertained in step 13.

15. Set the resistance load module for a resistance of 120 Ω.

16. Turn on the power supply and adjust the output voltage for a value of 120 V.

17. Measure and record the following values:

    E_P = ______________ V

    E_S = ______________ V

    I_P = ______________ A

    I_S = ______________ A

18. Compare these values with the computed values.

19. **Return the voltage to 0 V and turn off the power supply.**

20. Change the primary leads from 1 and 2 to 5 and 9. Also reset the value of the AC voltmeter connected to the primary to 100 V.

21. Change the secondary leads from 5 and 9 to 1 and 2, as in Figure 18-4. Also reset the value of the AC voltmeter connected to the secondary to 250 V.

22. Compute the turns ratio of the transformer by dividing the secondary voltage by the primary voltage. It is to be assumed that 60 V is to be connected to the primary winding.

    Ratio = ______________

    NOTE: The preceding transformer had a turns ratio of 2:1, which means that there are 2 turns of wire in the primary winding for every 1 turn of wire in the secondary winding. This transformer has a turns ratio of 1:2, which means that there is 1 turn of wire in the primary winding for every 2 turns in the secondary winding.

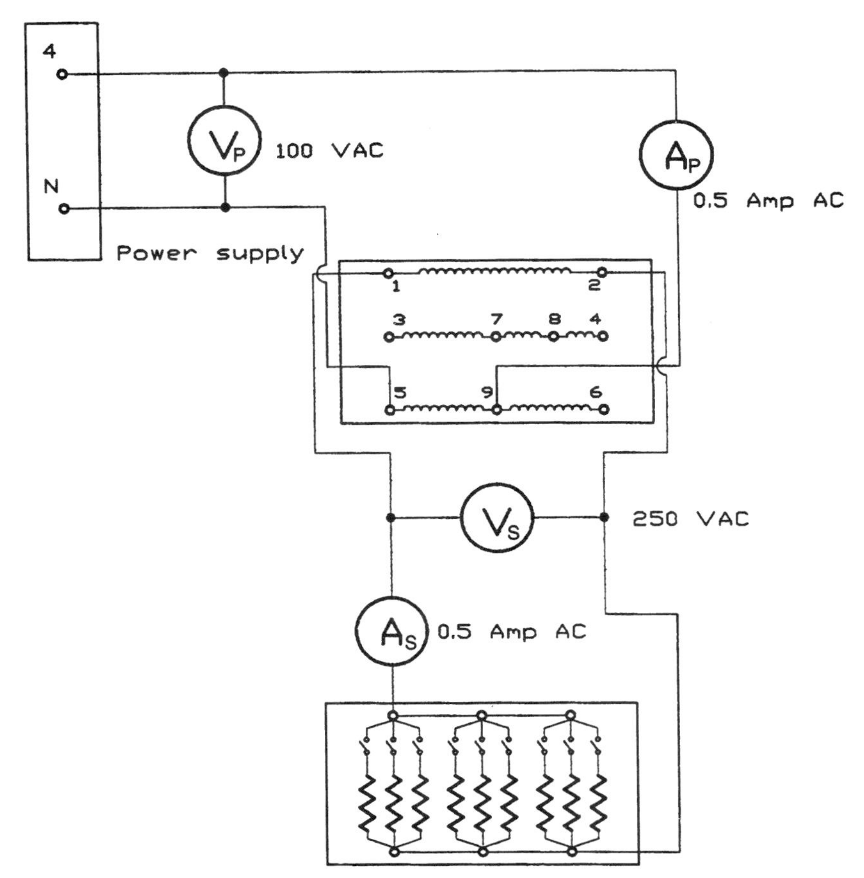

FIGURE 18-4 Changing the primary and secondary connections

23. Assume that a resistance of 600 Ω is to be connected to the secondary of this transformer. Compute the amount of current flow in this winding.

    I_S = ______________ A

24. Compute the amount of current flow in the primary winding using the formula

$$\frac{E_P}{E_S} = \frac{I_S}{I_P}$$

    I_P = ______________ A

25. Now compute the amount of current flow in the primary winding using the turns ratio.

    I_P = ______________ A

26. Set the resistance load module for a value of 600 Ω.
27. Turn on the power supply and adjust the output voltage for a value of **60 V.**
28. Measure and record the following values:

    E_P = ______________ V

    E_S = ______________ V

    I_P = ______________ A

    I_S = ______________ A

29. Compare these values with the computed values.

30. **Return the voltage to 0 V and turn off the power supply.**

31. Change the primary winding from terminals 5 and 9 to terminals 1 and 2. Also change the AC voltmeter to a setting of 250 V.

32. Change the secondary leads from terminals 1 and 2 to terminals 8 and 4. Also change the setting of the AC voltmeter connected to the secondary winding to 100 V, as in Figure 18-5.

33. Compute the turns ratio of this transformer.

    Ratio = ______________

34. Assume that a voltage of 120 V is to be connected to the primary winding and a load resistance of 85.7 Ω is to be connected to the secondary winding. Compute the amount of current flow in the secondary winding.

    I_S = ______________ A

35. Calculate the primary current using the formula

$$\frac{E_P}{E_S} = \frac{I_S}{I_P}$$

    I_P = ______________ A

36. Calculate the primary current using the turns ratio.

    I_P = ______________ A

37. Set the resistance load module for a value of 85.7 Ω.

38. Turn on the power supply and adjust the output voltage to a value of 120 V.

39. Measure and record the following values:

    E_P = ______________ V

    E_S = ______________ V

    I_P = ______________ A

    I_S = ______________ A

40. Compare the measured values with the computed values.

41. **Return the voltage to 0 V and turn off the power supply.**

42. Disconnect the circuit and return the components to their proper place.

Review Questions

1. All values of a transformer are proportional to the

   ______________ ______________.

2. What is excitation current?

   ____

3. A transformer has a primary voltage of 480 V and a secondary voltage of 24 V. What is the turns ratio of this transformer?

   ____

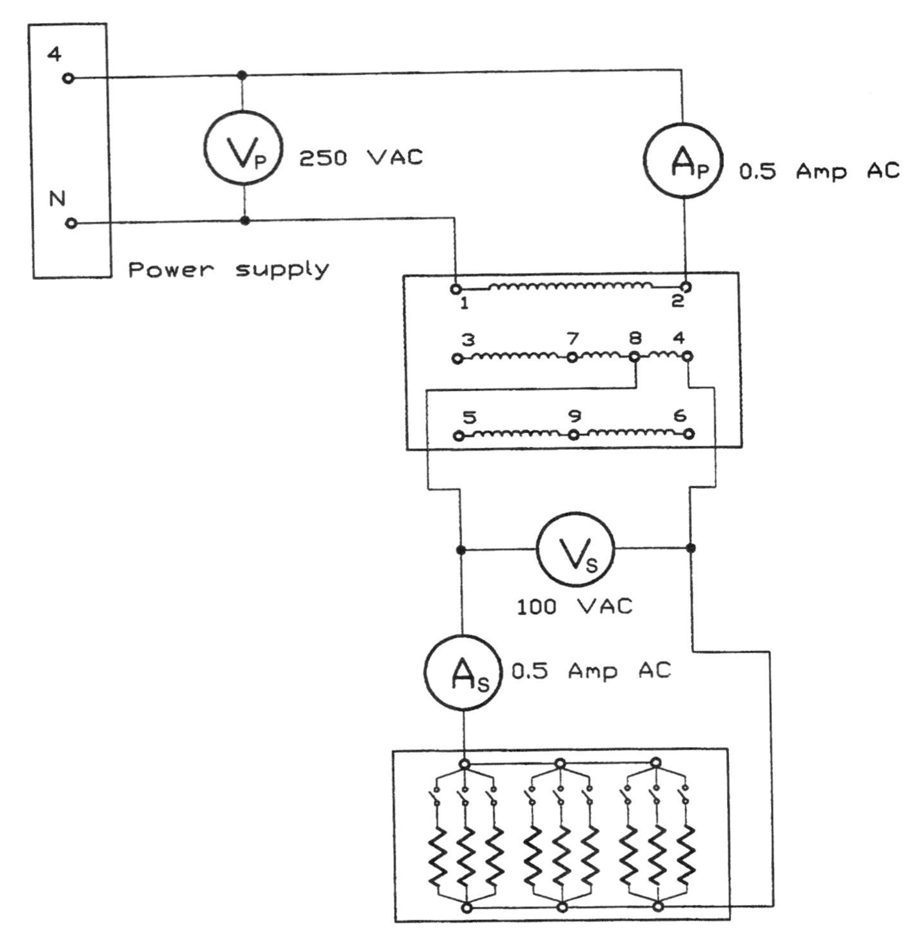

FIGURE 18-5 Reconnecting the transformer

4. A transformer has a turns ratio of 2.5:1. The primary current is 8 A. What is the secondary current?

   ______________ A

5. A transformer has a primary current of 0.5 A and a secondary current of 6 A. What is the turns ratio?

   ____

6. If the primary voltage of the transformer in question 5 is 120 V, what is the secondary voltage?

   ______________ V

7. A transformer has an excitation current of 0.2 A, a turns ratio of 1:4, and a primary voltage of 120 V. A load resistance of 240 Ω is connected to the secondary winding. What is the total primary current?

   ______________ A

Exercise 19

Transformer Polarities

Objectives

After completing this lab you should be able to:

- Discuss transformer polarity.
- Test transformer windings for polarity.
- Read and interpret polarity markings on a schematic.

Materials and Equipment

Power supply module	EMS 8821
AC voltmeter module	EMS 8426
Transformer module	EMS 8341

Discussion

To understand what is meant by transformer polarity, you must consider the voltage produced across a winding during some point in time. In a 60-Hz AC circuit, the voltage changes polarity 60 times per second. Transformer polarity, however, involves the relationship between the different windings at the same point in time. It will, therefore, be assumed that this point in time occurs when the peak positive voltage is being produced across the winding.

SCHEMATICS

It is common practice to indicate on a schematic diagram the polarity of the transformer windings by placing a dot beside one end of each winding as shown in Figure 19-1. These dots signify that the polarity is the same at that point in time for each winding. For example, assume the voltage applied to the primary winding is at its peak positive point at the terminal indicated by the dot. The voltage at the dotted lead of the secondary will be at its peak positive value at the same time.

This same type of polarity notation is used for transformers that have more than one primary or secondary winding. An example of a transformer with a multisecondary is shown in Figure 19-2.

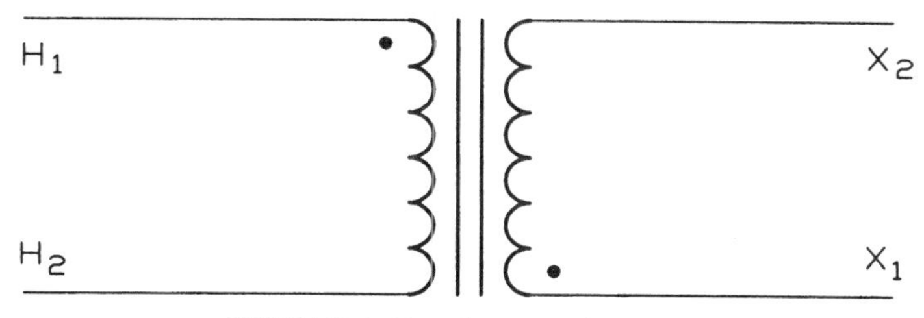

FIGURE 19-1 Transformer polarity dots

ADDITIVE AND SUBTRACTIVE POLARITIES

Transformer windings can be tested by connecting them as an autotransformer and testing for additive or subtractive polarity. (Additive polarity connections are often referred to as "boost," and subtractive polarity connections are often referred to as "buck.") This is done by connecting one lead of the secondary to one lead of the primary and measuring the voltage across both windings, as in Figure 19-3. The transformer shown in this example has a primary voltage rating of 120 V and a secondary voltage rating of 24 V. This same circuit has been redrawn in Figure 19-4 to show the connection more clearly. Notice that the secondary winding has been connected in series with the primary winding. The transformer now contains only one winding and is, therefore, an autotransformer. When 120 V is applied to the primary winding, the voltmeter connected across the secondary will indicate either the sum of the two voltages or the difference between the two voltages. If this voltmeter indicates 144 V (120 + 24 = 144) the windings are connected additive (boost), and polarity dots can be placed as shown in Figure 19-5. Notice that in this connection the secondary voltage is added to the primary voltage.

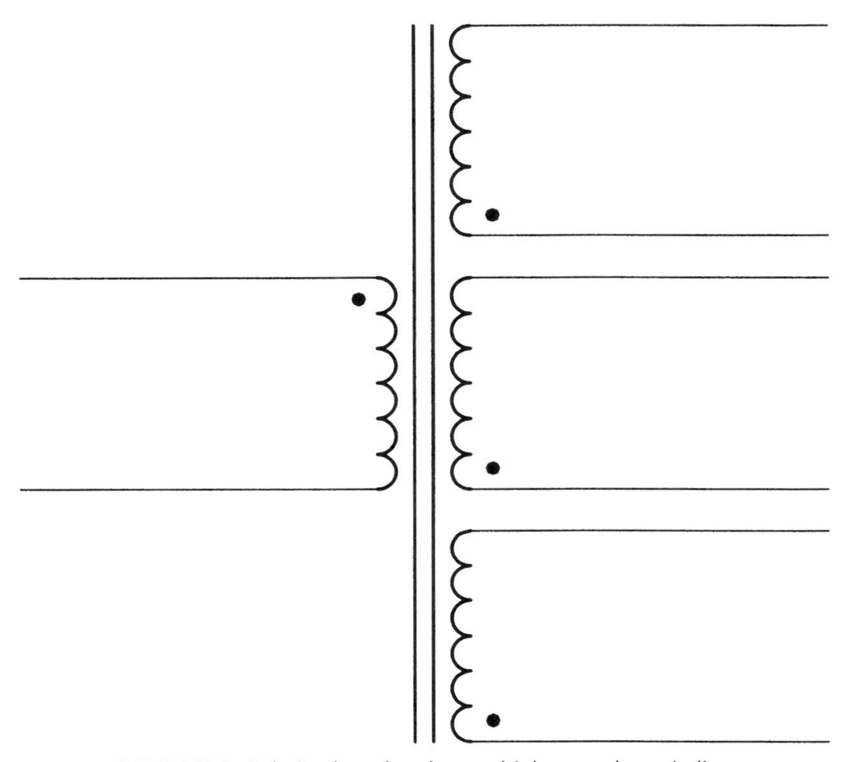

FIGURE 19-2 Polarity dots placed on multiple secondary windings

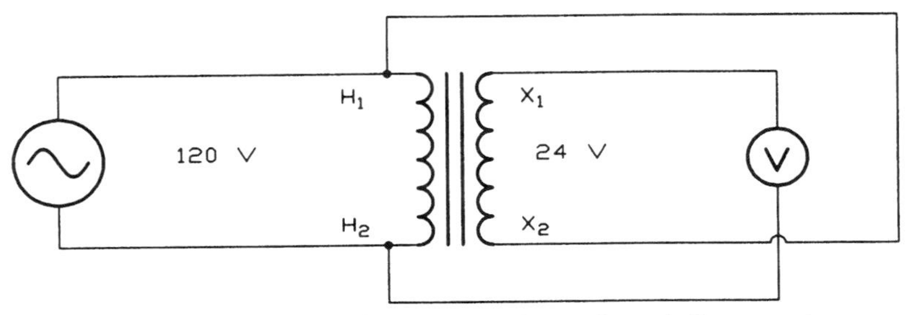

FIGURE 19-3 Connecting the primary and secondary windings together

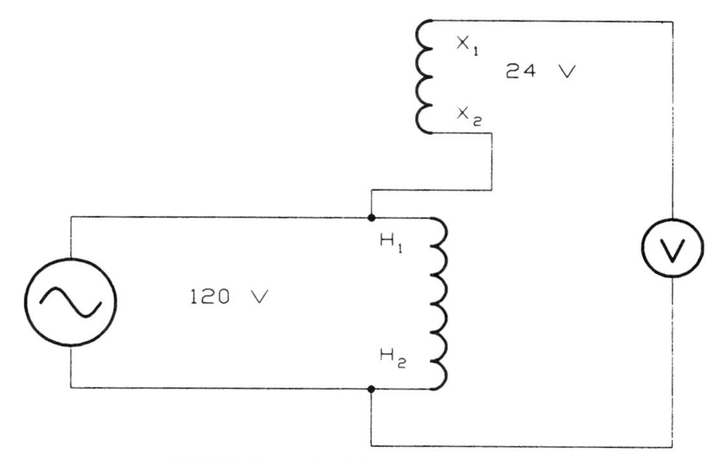

FIGURE 19-4 Simplifying the connection

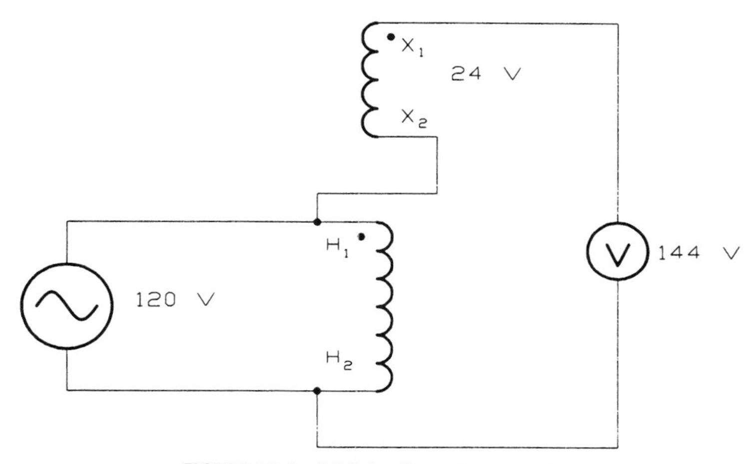

FIGURE 19-5 Additive (boost) connection

If the voltmeter connected across both windings, which is now the secondary winding, should indicate a voltage of 96 V (120 – 24 = 96) the windings are connected subtractive (buck), and polarity dots would be placed as shown in Figure 19-6.

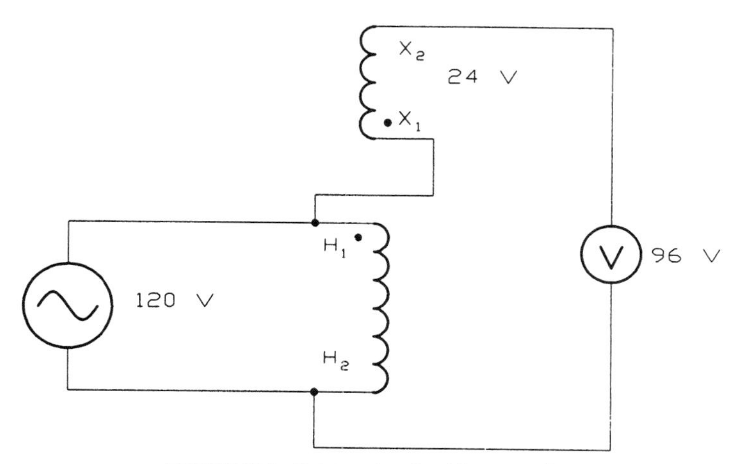

FIGURE 19-6 Subtractive (buck) connection

To help in the understanding of additive and subtractive polarity, arrows can be used to indicate a direction of greater-than or less-than values. This is the same principle as vector addition. In Figure 19-7, arrows have been added to indicate the direction in which the dot is to be placed. In this example, the transformer is connected additive, and both of the arrows point in the same direction. Notice that the arrows point to the dots. Figure 19-8 shows that values of the two arrows add to produce 144 V.

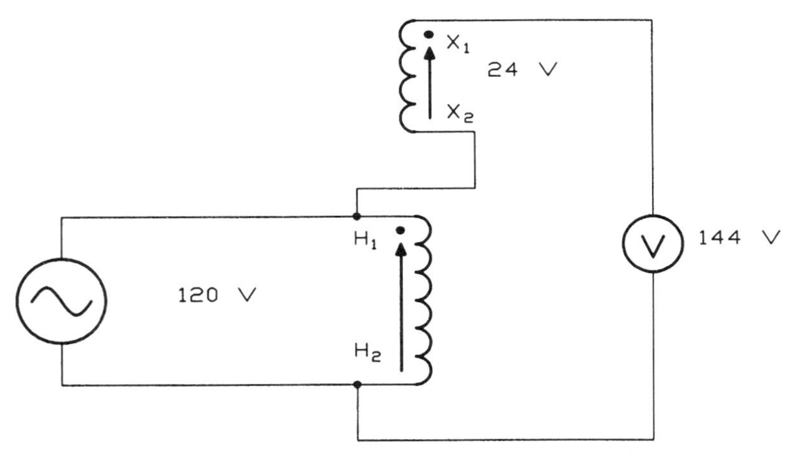

FIGURE 19-7 Arrows help show placement of dots for additive polarity connection.

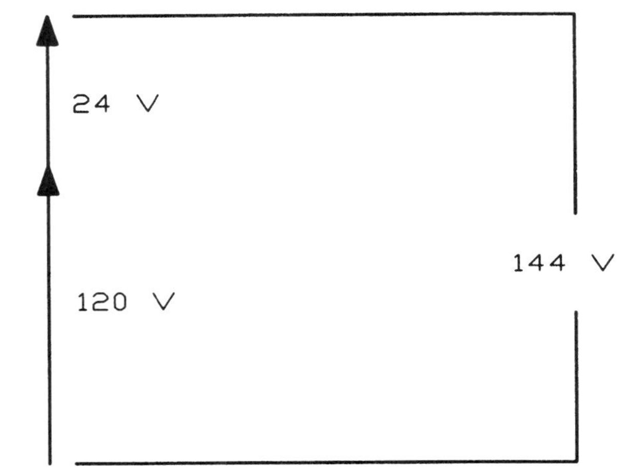

FIGURE 19-8 Arrows help illustrate addition of the two voltages.

In Figure 19-9, arrows have been added to a subtractive connection. In this instance, the arrows point in opposite directions, and the voltage of one tries to cancel the voltage of the other. The result is that the smaller value is eliminated and the larger value is reduced as shown in Figure 19-10.

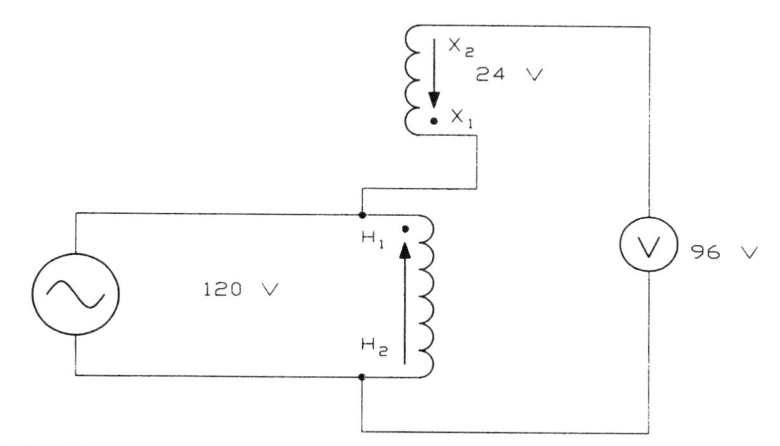

FIGURE 19-9 Arrows help indicate proper placement of dots for subtractive polarity.

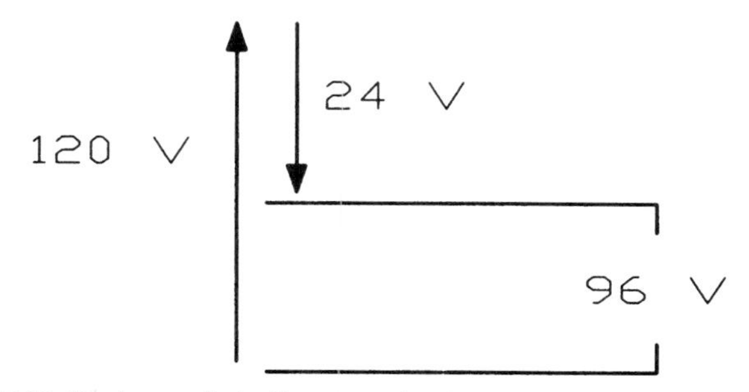

FIGURE 19-10 Arrows help illustrate why the two voltage values subtract.

Name____ Date________________

Procedure

1. Connect the circuit shown in Figure 19-11.
2. Connect terminal 5 of the transformer to terminal 2 of the transformer.
3. Connect a 250-VAC voltmeter across transformer terminals 1 and 6.
4. Turn on the power supply and increase the voltage applied to the primary to a value of 120 V.
5. Measure the voltage across terminals 1 and 6.

   ____________ V

6. Is this connection additive or subtractive?

   ____

7. If a polarity dot were placed beside terminal 2 of the transformer, would a second polarity dot be placed beside terminal 5 or terminal 6?

   ____

8. **Return the voltage to 0 V and turn off the power supply.**
9. Reconnect the transformer module so that terminal 6 is connected to terminal 2 and the voltmeter is connected across terminals 1 and 5.
10. Turn on the power supply and adjust the voltage for 120 V.
11. Measure the voltage between terminals 1 and 5.

    ____________ V

12. Is the connection additive or subtractive?

    ____

13. **Return the voltage to 0 V and turn off the power supply.**
14. Reconnect the transformer module so that terminal 3 is connected to terminal 2 and the 250-VAC voltmeter is connected across terminals 1 and 4.

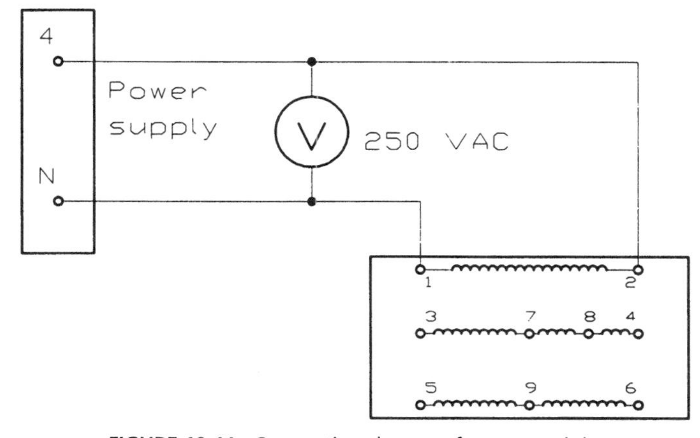

FIGURE 19-11 Connecting the transformer module

NOTE: In this example it will be necessary to reduce the voltage connected to the primary. The winding between terminals 1 and 2 is rated 120 V, and the winding between terminals 3 and 4 is rated 208 V. If these windings were connected in additive polarity, a voltage of 328 V would be produced. This voltage is greater than the full range of the voltmeter. For this reason, the voltage applied to the primary winding will be 60 V. This will produce a voltage of 104 V between terminals 3 and 4.

15. Turn on the power supply and adjust the output voltage for a value of **60 V.**
16. Measure the voltage between terminals 1 and 4.

    ____________ V
17. Are the windings connected additive or subtractive?

    ____
18. **Return the voltage to 0 V and turn off the power supply.**
19. If a polarity dot were placed beside terminal 2 on the transformer module, would a second polarity dot be placed beside terminal 3 or 4?

    ____
20. Reconnect the transformer module so that terminal 4 is connected to terminal 2 and the 250-VAC voltmeter is connected across terminals 1 and 3.
21. Turn on the power supply and adjust the output voltage for a value of **60 V.**
22. Measure the voltage between terminals 1 and 3.

    ____________ V
23. Are these windings connected additive or subtractive?

    ____
24. **Return the voltage to 0 V and turn off the power supply.**
25. Disconnect the circuit and return the components to their proper place.

Review Questions

1. What do the dots shown beside the terminal leads of a transformer represent on a schematic?

   ____
2. A transformer has a primary voltage rating of 240 V and a secondary voltage rating of 80 V. If the windings are connected subtractive, what voltage will appear across the entire connection?

   ____________ V
3. If the windings of the transformer in question 2 were connected additive, what voltage would appear across the entire winding?

   ____________ V
4. The primary leads of a transformer are labeled 1 and 2. The secondary leads are labeled 3 and 4. If polarity dots are placed beside leads 1 and 4, which secondary lead would be connected to terminal 2 to make the connection additive?

   ____

Exercise 20

The Autotransformer

Objectives

After completing this lab you should be able to:

- Discuss the operation of autotransformers.
- Compute the turns ratio of an autotransformer.
- Compute voltage and current values of an autotransformer.
- Connect an autotransformer and make measurements using test equipment.

Materials and Equipment

Power supply module	EMS 8821
Variable resistance module	EMS 8311
Transformer module	EMS 8341
AC ammeter module	EMS 8425
AC voltmeter module	EMS 8426

Discussion

The autotransformer is a transformer that contains only one winding. In this type of transformer the same winding is used as both the primary and secondary, as shown in Figure 20-1. Autotransformers have high efficiencies and are generally less expensive than other transformers. However, it should be noted that they do not provide line isolation between the power supply and the load.

VOLTAGE CALCULATIONS

The same formulas are used to compute the values of turns ratio, voltage, and current in an autotransformer as are used with an isolation type transformer. For example, assume that the primary of the transformer shown in Figure 20-1 has a total of 120 turns of wire and is connected to 120 VAC. Now assume that the secondary winding has a total of 80 turns of wire. The turns ratio of this transformer is 1.5:1 (120/80 = 1.5). Because the primary winding is connected to 120 V, the secondary voltage will be 80 V. This value of voltage can be computed using any one of three methods. One method is to use the formula shown below.

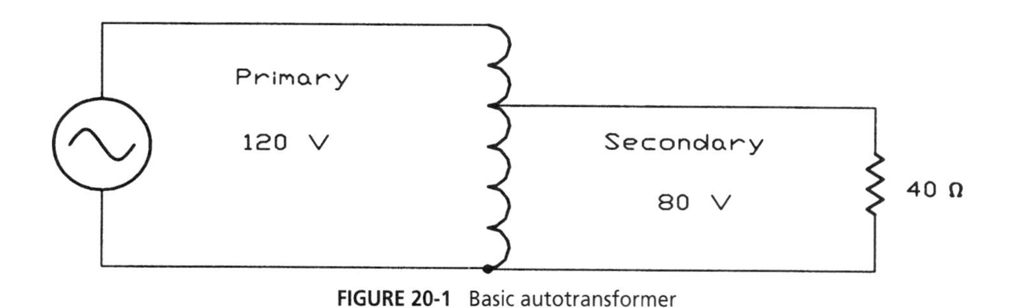

FIGURE 20-1 Basic autotransformer

$$\frac{E_P}{E_S} = \frac{N_P}{N_S}$$

$$\frac{120}{E_S} = \frac{120}{80}$$

$$120 \times E_S = 9600$$

$$E_S = 80 \text{ V}$$

The second method is to use the turns ratio. This transformer has a turns ratio of 1.5:1. This means that there are 1.5 turns of wire in the primary winding for every 1 turn of wire in the secondary. There are fewer turns of wire in the secondary, so the secondary voltage will be lower than the primary voltage. The secondary voltage can be computed by dividing the primary voltage by the turns ratio 120 V/1.5 = 80 V.

NOTE: If the turns ratio had been 1:1.5 it would have indicated that there are more turns of wire in the secondary. In that event the primary voltage would have been ***multiplied*** by the turns ratio.

The third method is to determine the volts per turn of the transformer. It was stated that this transformer has 120 turns of wire in its primary and 80 turns in its secondary. Because the primary is connected across 120 V, the primary winding has a ratio of 1 volt per turn of wire (120 V/120 turns = 1 V per turn). The voltage of the secondary can be computed by multiplying the volts per turn by the number of secondary turns.

$$80 \text{ turns} \times 1 \text{ V per turn} = 80 \text{ V}$$

CURRENT CALCULATIONS

The amount of primary and secondary current in the autotransformer can be computed using the same methods as those used for isolation type transformers. The transformer shown in Figure 20-1 has a load resistance of 40 Ω connected to the secondary. The amount of secondary current can be computed using Ohm's law.

$$I = \frac{E}{R}$$

$$I = \frac{80 \text{ V}}{40\ \Omega}$$

$$I = 20 \text{ A}$$

The amount of current flow through the primary winding can be computed using the formula

$$\frac{E_P}{E_S} = \frac{I_S}{I_P}$$

$$\frac{120 \text{ V}}{80 \text{ V}} = \frac{2}{I_P}$$

$$120 \times I_P = 160$$

$$I_P = 1.333 \text{ A}$$

The current can also be computed using the turns ratio. Because the primary voltage is greater than the secondary voltage, the primary current will be less than the secondary current.

$$\frac{2 \text{ A}}{1.5} = 1.333 \text{ A}$$

Name ______________________________ Date ______________

Procedure

1. Connect the circuit shown in Figure 20-2.
2. Assume that a voltage of 120 V is to be supplied to transformer terminals 5 and 6. The transformer secondary will be across terminals 6 and 9. What will be the secondary voltage?

   E_S = ______________ V
3. Compute the turns ratio of this transformer.

   Ratio = ______________
4. Set the resistance module for a value of 120 Ω.
5. Compute the amount of secondary current that should flow in this circuit using Ohm's law.

   I_S = ______________ A
6. Using the formula shown, compute the amount of primary current that should flow in this circuit.

$$\frac{E_P}{E_S} = \frac{I_S}{I_P}$$

   I_P = ______________ A
7. Compute the amount of primary current that should flow using the turns ratio.

   I_P = ______________ A

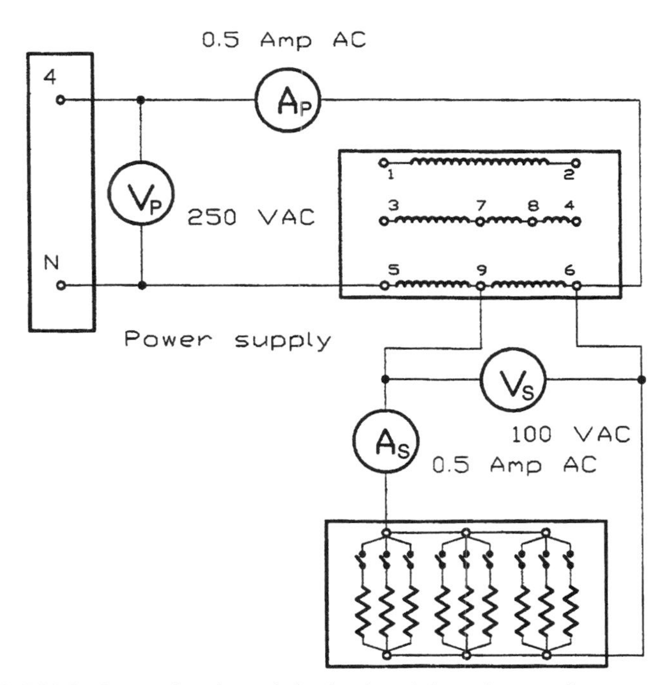

FIGURE 20-2 Connecting the resistive load module to the transformer module

8. Turn on the power supply and adjust the output voltage for a value of 120 V.
9. Measure and record the values shown below.

   E_S = ______________ V

   I_S = ______________ A

   I_P = ______________ A

10. Compare these values with the computed values.
11. **Return the voltage to 0 V and turn off the power supply.**
12. Reconnect the circuit as shown in Figure 20-3. Notice that terminals 3 and 4 on the transformer module are now used as the primary and terminals 7 and 8 are used as the secondary. Also notice that the primary of the transformer is now connected to terminals 4 and 5 of the power supply. Assuming that 208 V is supplied to the primary winding of the transformer, what will be secondary voltage?

   E_S = ______________ V

13. Compute the turns ratio of the transformer.

   Ratio = ______________

14. Set the resistance module for a value of 300 Ω.
15. Compute the amount of current that should flow in the secondary circuit using the formula shown below.

$$I = \frac{E}{R}$$

   I_S = ______________ A

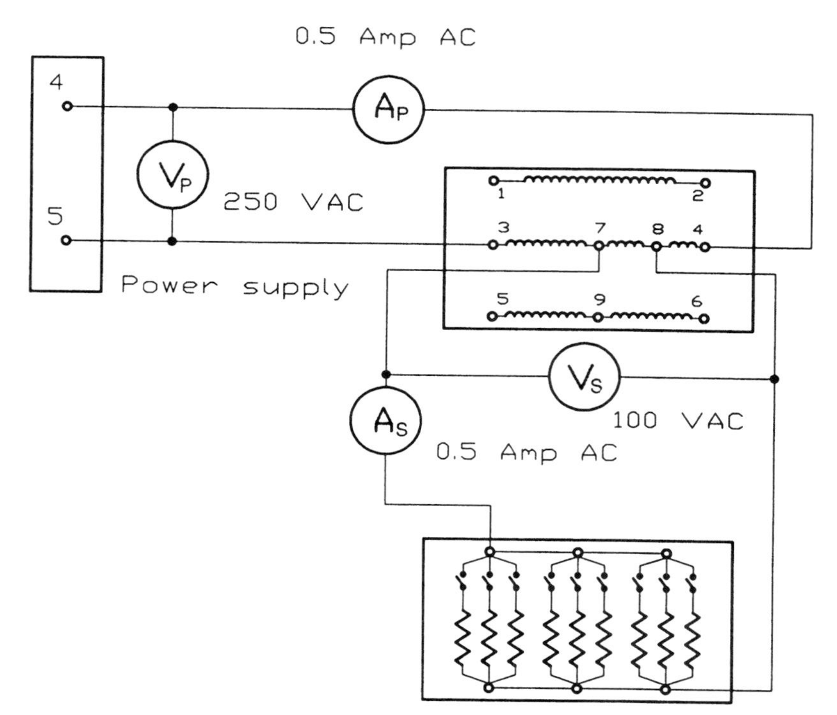

FIGURE 20-3 Connection of the second circuit

16. Compute the amount of primary current using the formula shown below.

$$\frac{E_P}{E_S} = \frac{I_S}{I_P}$$

I_P = ____________ A

17. Compute the amount of primary current using the turns ratio.

I_P = ____________ A

18. Turn on the power supply and adjust the output voltage for a value of 208 V.
19. Measure and record the following values:

E_S = ____________ V

I_S = ____________ A

I_P = ____________ A

20. Compare these values with the computed values.
21. **Return the voltage to 0 V and turn off the power supply.**
22. Disconnect the circuit and return the components to their proper place.

Review Questions

1. Name two advantages of the autotransformer.

A. ____

B. ____

2. Name a disadvantage of an autotransformer.

____

3. To answer the following questions, refer to the transformer module.

A. Assume that a voltage of 60 V is connected to terminals 5 and 9. If terminals 5 and 6 are used as the secondary, what will be the output voltage?

____________ V

B. What is the turns ratio of this connection?

____

C. Assume that a resistance of 480 Ω has been connected across terminals 5 and 6. What would be the current flow through the secondary circuit?

____________ A

D. What would be the current flow in the primary circuit?

____________ A

Exercise 21

Three-Phase Transformers

Objectives

After completing this lab you should be able to:

- Discuss the operation of three-phase transformers.
- Connect three single-phase transformers to form a three-phase bank.
- Calculate voltage values for a three-phase transformer connection.

Materials and Equipment

Power supply module	EMS 8821
Transformer module (3)	EMS 8341
AC ammeter module	EMS 8425
AC voltmeter module	EMS 8426
Resistance load module	EMS 8311

Discussion

A three-phase transformer is constructed by winding three single-phase transformers on a single core. If three-phase transformation is needed and a three-phase transformer of the proper size and turns ratio is not available, three single-phase transformers can be connected to form a three-phase bank.

Three-phase transformers are connected in delta or wye configurations. A wye-delta transformer, for example, has its primary winding connected in a wye and its secondary winding connected in a delta, as shown in Figure 21-1. When three single-phase transformers are used to make a three-phase transformer bank, their primary and secondary windings are connected in a wye or delta connection. The windings in Figure 21-1 have been labeled A, B, and C. One end of each primary lead has been labeled H_1 and the other end has been labeled H_2. One end of each secondary lead has been labeled X_1 and the other end has been labeled X_2.

Figure 21-2 shows three single-phase transformers. The primary leads of each transformer have been labeled H_1 and H_2, and the secondary leads have been labeled X_1 and X_2. Notice also that each transformer has been designated as A, B, or C.

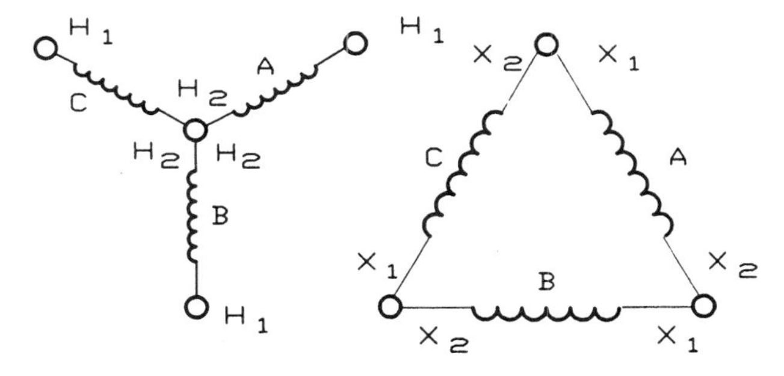

FIGURE 21-1 Wye-delta transformer connection

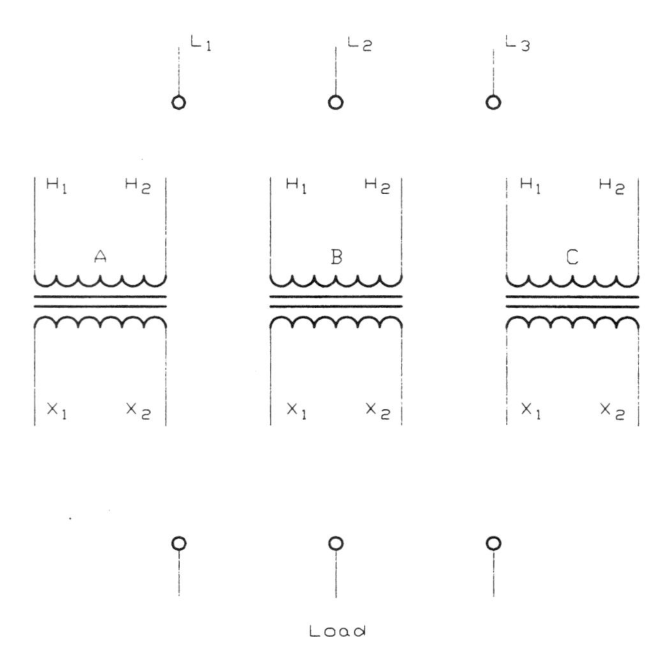

FIGURE 21-2 Three single-phase transformers

TRANSFORMER CONNECTION

The schematic diagram of Figure 21-1 will be used to connect the three single-phase transformers into a three-phase, wye-delta connection as shown in Figure 21-3.

The primary winding will be tied into a wye connection first. The schematic in Figure 21-1 shows that the H_2 leads of the three primary windings are connected together, and the H_1 lead of each winding is open for connection to the incoming power line. Notice in Figure 21-3 that the H_2 leads of the three primary windings have

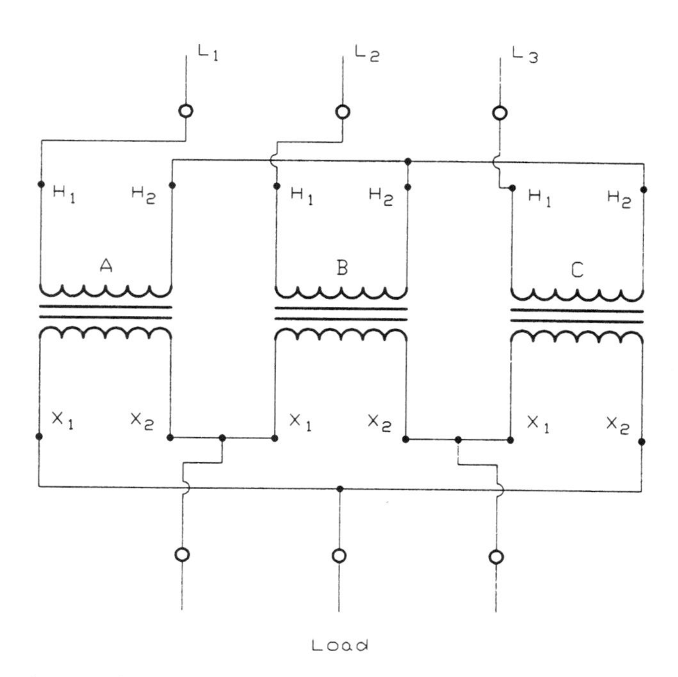

FIGURE 21-3 Three single-phase transformers connected to form a wye-delta three-phase bank

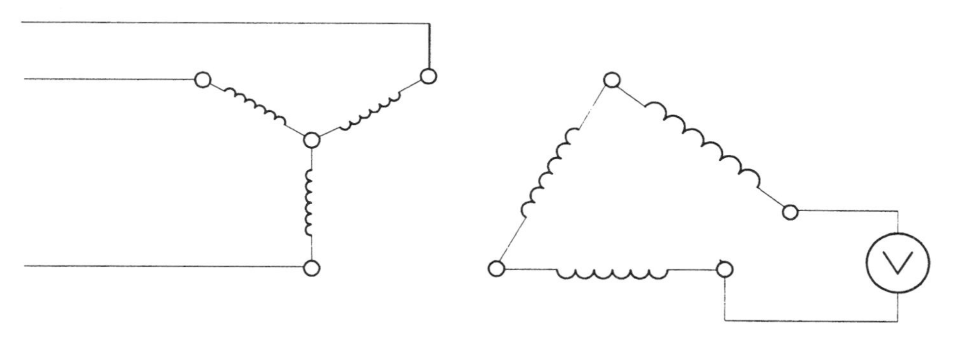

FIGURE 21-4 A delta connection should be tested for proper polarity before final connection is made.

been connected together, and the H_1 lead of each winding has been connected to the incoming power line.

Figure 21-1 shows that the X_1 lead of transformer A is connected to the X_2 lead of transformer C. Notice that this same connection has been made in Figure 21-3. The X_1 lead of transformer B is connected to the X_2 lead of transformer A, and the X_1 lead of transformer C is connected to the X_2 lead of transformer B. The load is connected to the points of the delta connection.

CLOSING A DELTA

Delta connections should be checked for proper polarity before making the final connection and applying power. If the phase winding of one transformer is reversed, an extremely high current will flow when power is applied. Proper phasing can be checked with a voltmeter as shown in Figure 21-4. If power is applied to the transformer bank before the delta connection is closed, the normal line voltage should be indicated by the voltmeter. If one phase winding has been reversed, however, the voltmeter will indicate double the amount of voltage. For example, assume that the output voltage of a delta secondary is 240 V. If the voltage is checked before the delta is closed, the voltmeter should indicate a voltage of 240 V if all windings have been phased properly. If one winding has been reversed, however, the voltmeter will indicate a voltage of 480 V.

THREE-PHASE TRANSFORMER CALCULATIONS

The formulas used for transformer calculations and three-phase calculations must be followed to compute the values of voltage and current for three-phase transformers. Another very important rule must be understood: ***Only phase values of voltage and current can be used to compute transformer values***. Refer to transformer A in Figure 21-2. All transformation of voltage and current takes place between the primary and secondary windings. Because these windings form the phase values of the three-phase connection, only phase and not line values can be used when calculating transformed voltages and currents.

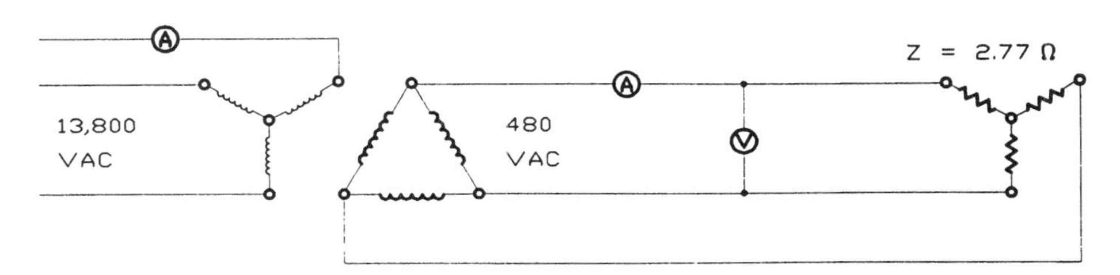

FIGURE 21-5 Wye-delta transformer bank connected to a wye load

A three-phase transformer connection is shown in Figure 21-5. Three single-phase transformers have been connected to form a wye-delta bank. The primary is connected to a three-phase line of 13,800 V, and the secondary voltage is 480 V. A three-phase resistive load with an impedance of 2.77 Ω per phase is connected to the secondary of the transformer. The following values will be computed for this circuit:

Phase voltage of the primary
Phase voltage of the secondary
Turns ratio of the transformer
Phase voltage of the load bank
Phase current of the load bank
Secondary line current
Phase current of the secondary
Phase current of the primary
Line current of the primary

The primary windings of the three single-phase transformers have been connected to form a wye connection. In a wye connection, the phase voltage is less than the line voltage by a factor of 1.732 (the square root of 3). Therefore, the phase value of voltage can be computed using the formula

$$E_{PHASE} = \frac{E_{LINE}}{1.732}$$
$$E_{PHASE} = \frac{13,800\ V}{1.732}$$
$$E_{PHASE} = 7967.67\ V$$

The secondary windings are connected as a delta. In a delta connection, the phase voltage and line voltage are the same.

$$E_{PHASE} = E_{LINE}$$
$$E_{PHASE} = 480\ V$$

The turns ratio can be computed by comparing the phase voltage of the primary with the phase voltage of the secondary.

$$\text{Turns ratio} = \frac{\text{primary voltage}}{\text{secondary voltage}}$$
$$\text{Turns ratio} = 16.6:1$$

The load bank is connected in a wye connection. The voltage across the phase of the load bank will be less than the line voltage by a factor of 1.732.

$$E_{PHASE} = \frac{E_{LINE}}{1.732}$$
$$E_{PHASE} = \frac{480\ V}{1.732}$$
$$E_{PHASE} = 277\ V$$

Now that the voltage across each of the load resistors is known, the current flow through the phase of the load can be computed using Ohm's law.

$$I = \frac{E}{R}$$
$$I = \frac{277\ V}{2.77\ \Omega}$$
$$I = 100\ A$$

Because the load is connected as a wye connection, the line current will be the same as the phase current.

$$I_{LINE} = 100 \text{ A}$$

The secondary of the transformer bank is connected as a delta. The phase current of the delta is less than the line current by a factor of 1.732.

$$I_{PHASE} = \frac{I_{LINE}}{1.732}$$

$$I_{PHASE} = \frac{100 \text{ A}}{1.732}$$

$$I_{PHASE} = 57.74 \text{ A}$$

The amount of current flow through the primary can be computed using the turns ratio. Because the primary has a higher voltage, it will have a lower current (volts × amps input must equal volts × amps output).

$$\text{primary current} = \frac{\text{secondary current}}{\text{turns ratio}}$$

$$I_{primary} = \frac{57.74 \text{ A}}{16.6}$$

$$I_{primary} = 3.48 \text{ A}$$

Because all transformed values of voltage and current take place across the phases, the primary has a phase current of 3.48 A. In a wye connection, the phase current is the same as the line current.

$$I_{LINE} = 3.48 \text{ A}$$

Name ______________________________ Date ______________

Procedure

1. Using the three EMS 8341 transformer modules, connect the circuit shown in Figure 21-6. It will be assumed that the input voltage at power supply terminals 4, 5, and 6 will be 208 V. It will also be assumed that the load resistors will have a value of 240 Ω per phase.
2. Compute the amount of phase voltage across each of the primary windings.

$$E_{PHASE} = \frac{E_{LINE}}{1.732}$$

$E_{PHASE(pri)}$ = ______________ V

3. What will be the phase voltage across each of the secondary windings? (This value is shown on the front of the transformer module.)

$E_{PHASE(sec)}$ = ______________ V

4. Using the values of phase voltage for the primary and secondary, calculate the turns ratio of the transformer.

$$\text{Ratio} = \frac{E_{PHASE(pri)}}{E_{PHASE(sec)}}$$

Turns ratio = ______________

5. Compute the amount of line voltage of the secondary.

$$E_{LINE} = E_{PHASE} \times 1.732$$

$E_{LINE(sec)}$ = ______________ V

6. Compute the amount of voltage across each of the resistors.

$$E_{PHASE} = \frac{E_{LINE}}{1.732}$$

$E_{PHASE(load)}$ = ______________ V

7. Using Ohm's law, calculate the amount of current flow through each phase of the load.

$$I_{PHASE} = \frac{E_{PHASE}}{R}$$

$I_{PHASE(load)}$ = ______________ A

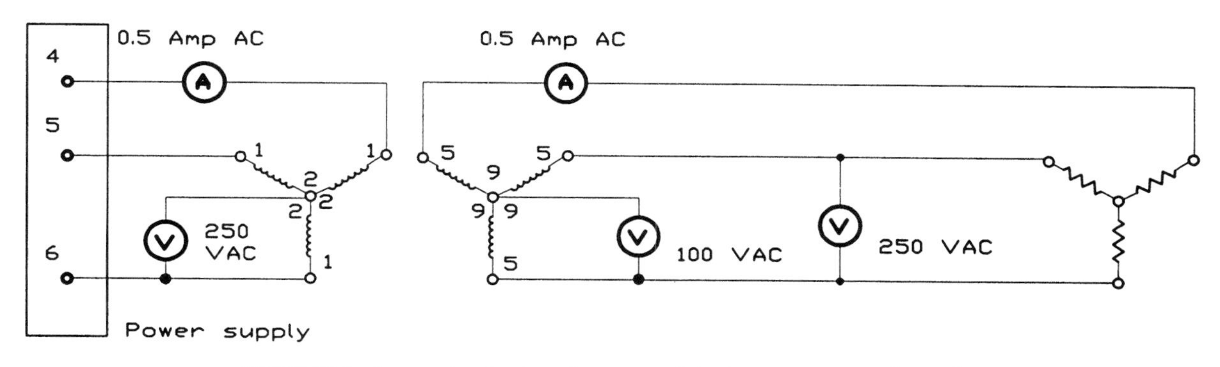

FIGURE 21-6 Connecting the transformer modules in a wye-wye configuration

8. Calculate the amount of line current that should be indicated by the AC ammeter.

$$I_{LINE} = I_{PHASE} \text{ in a wye connection}$$

I_{LINE} = ______________ A

9. Compute the amount of current flow through the phase of the secondary winding.

$$I_{PHASE} = I_{LINE} \text{ in a wye connection}$$

$I_{PHASE(sec)}$ = ______________ A

10. Using the turns ratio, calculate the amount of current flow through the phase of the primary.

$$I_{pri} = \frac{I_{sec}}{ratio}$$

$I_{PHASE(pri)}$ = ______________ A

11. Calculate the amount of line current that should be indicated by the AC ammeter connected to the primary of the transformer.

$$I_{LINE} = I_{PHASE} \text{ in a wye connection}$$

$I_{LINE(pri)}$ = ______________ A

12. Set the resistance load module to have a resistance of 240 Ω on each phase.
13. Turn on the power supply and adjust the output voltage for a value of 208 V.
14. Measure and record the following values:

$E_{PHASE(pri)}$ = ______________ V

$E_{PHASE(sec)}$ = ______________ V

$E_{LINE(sec)}$ = ______________ V

$I_{LINE(sec)}$ = ______________ A

$I_{LINE(pri)}$ = ______________ A

15. Compare these measured values with the computed values.
16. **Return the voltage to 0 V and turn off the power supply.**
17. Reconnect the circuit as shown in Figure 21-7. In this circuit it will be assumed that the input voltage from terminals 4, 5, and 6 of the power supply is 208 V, and the load resistance will have a value of 400 Ω per phase.
18. Compute the phase voltage of the primary.

$$E_{PHASE} = E_{LINE} \text{ in a delta connection}$$

$E_{PHASE(pri)}$ = ______________ V

19. List the phase voltage of the secondary. (This value can be found on the front of the transformer module.)

$E_{PHASE(sec)}$ = ______________ V

20. Using the phase value of voltage for the primary and secondary, calculate the turns ratio of the transformer.

$$\text{Turns ratio} = \frac{E_{PHASE(pri)}}{E_{PHASE(sec)}}$$

Turns ratio = ______________

21. Calculate the value of line voltage for the secondary of the transformer.

$$E_{LINE} = E_{PHASE} \times 1.732$$

$E_{LINE(sec)}$ = ______________ V

22. Compute the phase value of voltage for the three-phase load.

$$E_{PHASE} = \frac{E_{LINE}}{1.732}$$

$E_{PHASE(load)}$ = ______________ V

23. Using the phase value of voltage and the resistance, calculate the phase current of the load.

$$I_{PHASE} = \frac{E_{PHASE}}{R}$$

$I_{PHASE(load)}$ = ______________ A

24. Calculate the amount of line current that should be indicated by the ammeter connected in the secondary circuit.

$$I_{LINE} = I_{PHASE} \text{ in a wye connection}$$

$I_{LINE(sec)}$ = ______________ A

25. Compute the amount of phase current in the secondary of the transformer.

$$I_{PHASE} = I_{LINE} \text{ in a wye connection}$$

$I_{PHASE(sec)}$ = ______________ A

26. Using the phase current of the secondary winding and the turns ratio, calculate the amount of phase current that should flow in the primary winding.

$$I_{(pri)} = \frac{I_{(sec)}}{\text{ratio}}$$

$I_{PHASE(pri)}$ = ______________ A

27. Compute the amount of line current that should be indicated by the ammeter connected to the primary of the transformer.

$$I_{LINE} = I_{PHASE} \times 1.732$$

$I_{LINE(pri)}$ = ______________ A

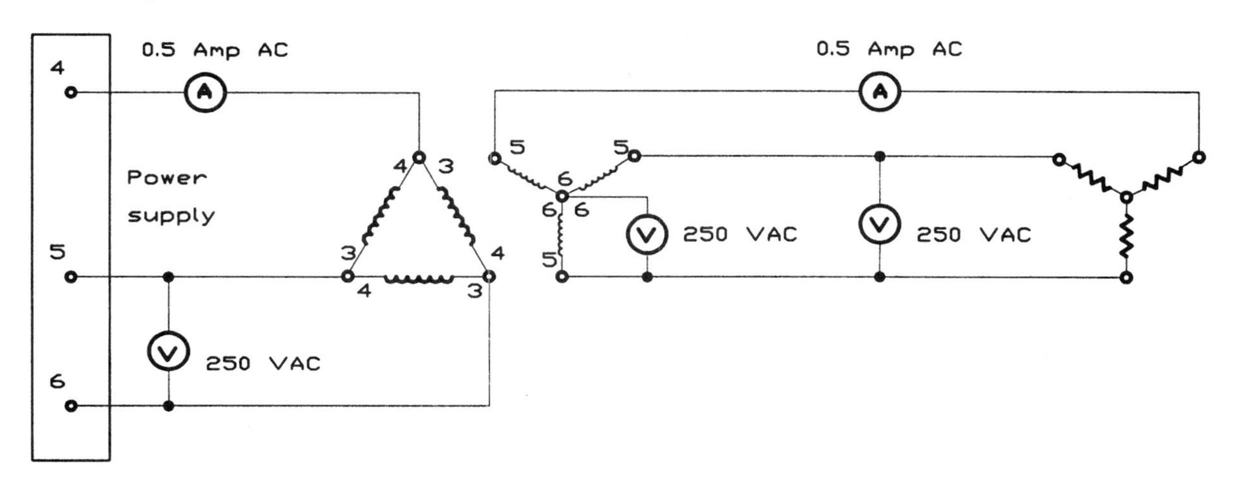

FIGURE 21-7 Connecting the transformer modules to form a delta-wye connection

28. Set the resistance load module for a resistance of 400 Ω on each phase.
29. Turn on the power supply and adjust the output voltage for a value of 208 V.
30. Measure and record the values listed below.

$E_{LINE(pri)}$ = ______________ V

$E_{LINE(sec)}$ = ______________ V

$E_{PHASE(sec)}$ = ______________ V

$I_{LINE(pri)}$ = ______________ A

$I_{LINE(sec)}$ = ______________ A

31. Compare these values with the computed values.
32. **Return the voltage to 0 V and turn off the power supply.**
33. Reconnect the circuit as shown in Figure 21-8. Assume that the primary of the transformer has an applied voltage of 208 V and the resistance bank has a value of 200 Ω per phase.
34. Compute the values of voltage and current listed below.

$$E_{PHASE} = \frac{E_{LINE}}{1.732}$$

$E_{PHASE(pri)}$ = ______________ V

$E_{PHASE(sec)}$ (Shown on module) = ______________ V

$$\frac{E_{PHASE(pri)}}{E_{PHASE(sec)}}$$

Turns ratio = ______________

$$E_{LINE} = E_{PHASE}$$

$E_{LINE(sec)}$ = ______________ V

$$E_{PHASE} = \frac{E_{LINE}}{1.732}$$

$E_{PHASE(load)}$ = ______________ V

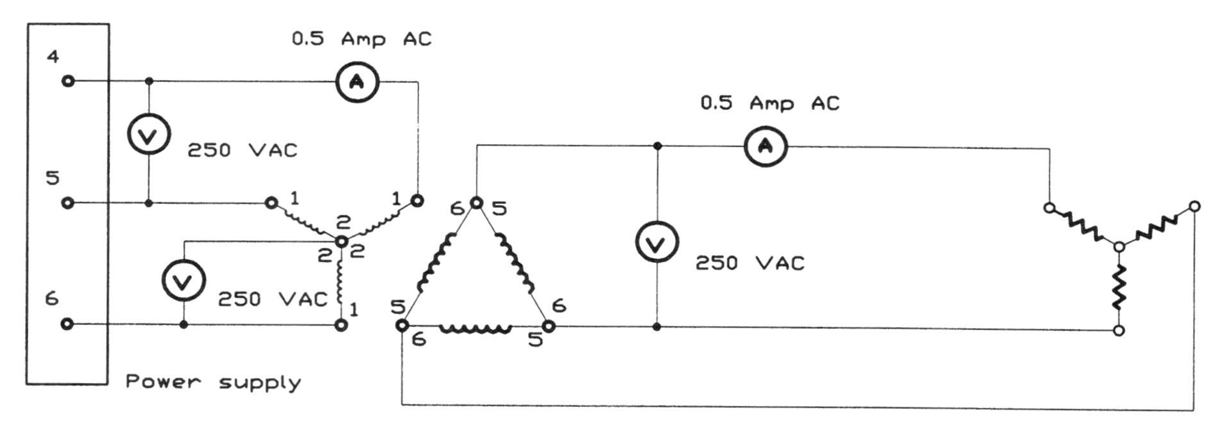

FIGURE 21-8 Second transformer connection

$$I_{PHASE} = \frac{E_{PHASE}}{R}$$

$I_{PHASE(load)}$ = ______________ A

$$I_{LINE} = I_{PHASE}$$

$I_{LINE(sec)}$ = ______________ A

$$I_{PHASE} = \frac{I_{LINE}}{1.732}$$

$I_{PHASE(sec)}$ = ______________ A

$$\frac{I_{PHASE(sec)}}{ratio}$$

$I_{PHASE(pri)}$ = ______________ A

$$I_{LINE} = I_{PHASE}$$

$I_{LINE(pri)}$ = ______________ A

35. Set the resistance load module for a value of 200 Ω per phase.
36. Turn on the power supply and adjust the output voltage for a value of 208 V.
37. Measure and record the values shown below.

$E_{LINE(pri)}$ = ______________ V

$E_{PHASE(pri)}$ = ______________ V

$E_{LINE(sec)}$ = ______________ V

$I_{LINE(sec)}$ = ______________ A

$I_{LINE(pri)}$ = ______________ A

38. Compare these values with the measured values.
39. **Return the voltage to 0 V and turn off the power supply.**
40. Reconnect the circuit as shown in Figure 21-9. Assume an applied voltage at the primary of 208 V and a resistance value of 171.4 Ω per phase.

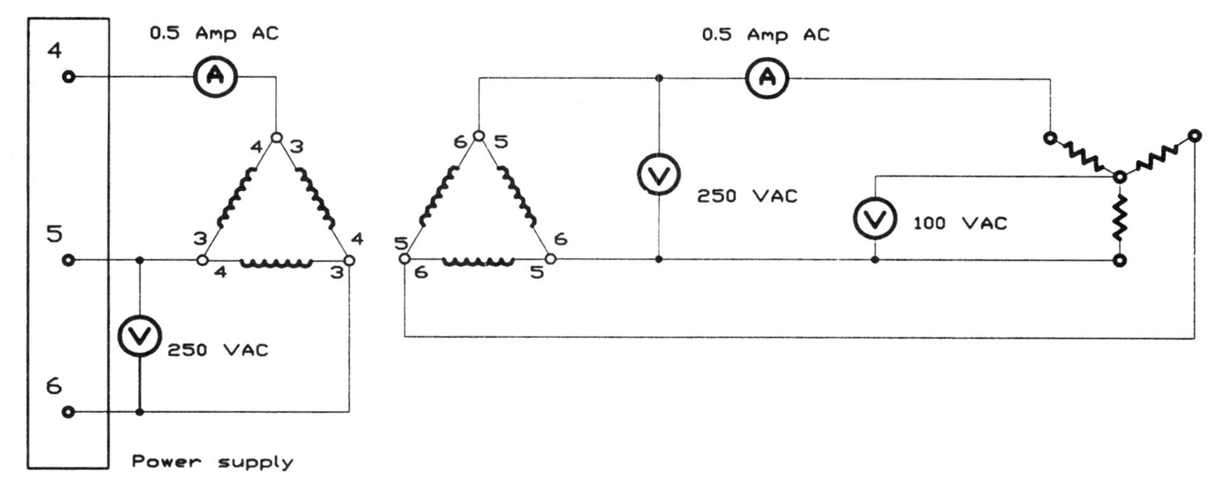

FIGURE 21-9 Connecting the transformer modules in a delta-delta configuration

41. Compute the values shown below.

$E_{PHASE} = E_{LINE}$

$E_{PHASE(pri)}$ = ______________ V

$E_{PHASE(sec)}$ (Found on transformer) = ______________ V

$\frac{E_{PHASE(pri)}}{E_{PHASE(sec)}}$

Turns ratio = ______________

$E_{LINE} = E_{PHASE}$

$E_{LINE(sec)}$ = ______________ V

$E_{PHASE} = \frac{E_{LINE}}{1.732}$

$E_{PHASE(load)}$ = ______________ V

$I_{PHASE} = \frac{E_{PHASE}}{R}$

$I_{PHASE(load)}$ = ______________ A

$I_{LINE} = I_{PHASE}$

$I_{LINE(sec)}$ = ______________ A

$I_{PHASE} = \frac{I_{LINE}}{1.732}$

$I_{PHASE(sec)}$ = ______________ A

$\frac{I_{PHASE(sec)}}{\text{ratio}}$

$I_{PHASE(pri)}$ = ______________ A

$I_{PHASE} \times 1.732$

$I_{LINE(pri)}$ = ______________ A

42. Set the load resistance for a value of 171.4 Ω per phase.
43. Turn on the power supply and adjust the output voltage for a value of 208 V.
44. Measure and record the values of voltage and current listed below.

$E_{LINE(pri)}$ = ______________ V

$E_{LINE(sec)}$ = ______________ V

$E_{PHASE(load)}$ = ______________ V

$I_{LINE(sec)}$ = ______________ A

$I_{LINE(pri)}$ = ______________ A

45. Compare these values with the computed values.
46. **Return the voltage to 0 V and turn off the power supply.**
47. Disconnect the circuit and return the components to their proper place.

Review Questions

1. What is a three-phase transformer?

   ____

2. Three single-phase transformers are connected into a three-phase bank. The primary is connected as a wye, and the secondary is connected as a delta. If 4160 V is connected to the primary, what is the phase voltage?

   E_{PHASE} = ____________ V

3. If the transformer has a turns ratio of 9.45:1, what is the phase voltage of the secondary?

   E_{PHASE} = ____________ V

4. If the secondary of the transformer bank in question 2 has a line current of 1200 A, what will be the phase current?

   I_{PHASE} = ____________ A

5. What will be the phase current of the primary of this transformer bank?

   I_{PHASE} = ____________ A

6. What will be the primary line current?

   I_{LINE} = ____________ A

Exercise 22

The Open Delta Connection

Objectives

After completing this lab you should be able to:

- Discuss the operation of an open delta transformer.
- Calculate values of voltage and current for open delta transformers.
- Connect an open delta transformer and make measurements using test instruments.

Materials and Equipment

Power supply module	EMS 8821
Variable-resistance module	EMS 8311
Transformer module (2)	EMS 8341
AC ammeter module	EMS 8425
AC voltmeter module	EMS 8426

Discussion

The open delta transformer connection can be made with only two transformers instead of three, as shown in Figure 22-1. This connection is often used when the amount of three-phase power needed is not excessive, such as in a small business. It should be noted that the output power of an open delta connection is only 86.6% of the rated power of the two transformers. For example, assume that two transformers, each having a capacity of 25 KVA (kilo-volt-amperes), are connected in an open delta connection. The total output power of this connection is 43.3 KVA ([25 + 25] × 0.866 = 43.3). This is sometimes stated as 57.7% (1/1.732) of a closed delta transformer bank. A closed delta bank would contain three transformers instead of two. Three transformers rated at 25 KVA each would have a total capacity of 75 KVA. 57.7% of 75 KVA = 43.3 KVA also (75 KVA × 0.577 = 43.3 KVA).

Another very common connection is shown in Figure 22-2. In this connection, one transformer is used to supply both three-phase and single-phase power while the second transformer supplies only its part of the three-phase power. The output voltage of this connection is 240 V three-phase. The larger rated transformer has been center-tapped to supply 120 V single-phase loads. Notice that transformer X has been rated at 25 KVA and transformer Y has been rated at 15 KVA.

Any delta-connected transformer bank that has one transformer center-tapped to supply single-phase loads has a high leg. In the circuit shown in Figure 22-2, output A

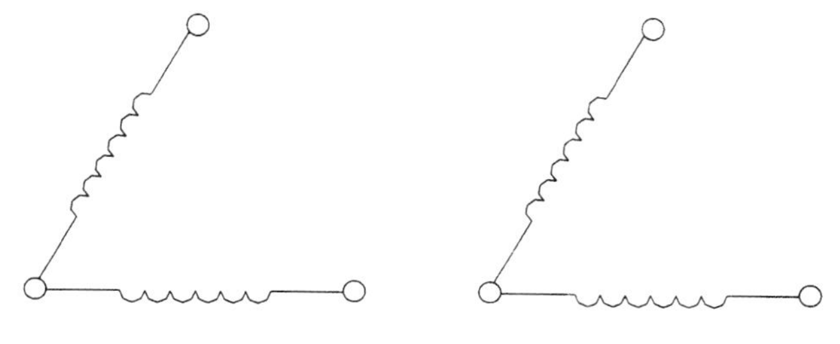

FIGURE 22-1 Open delta connection

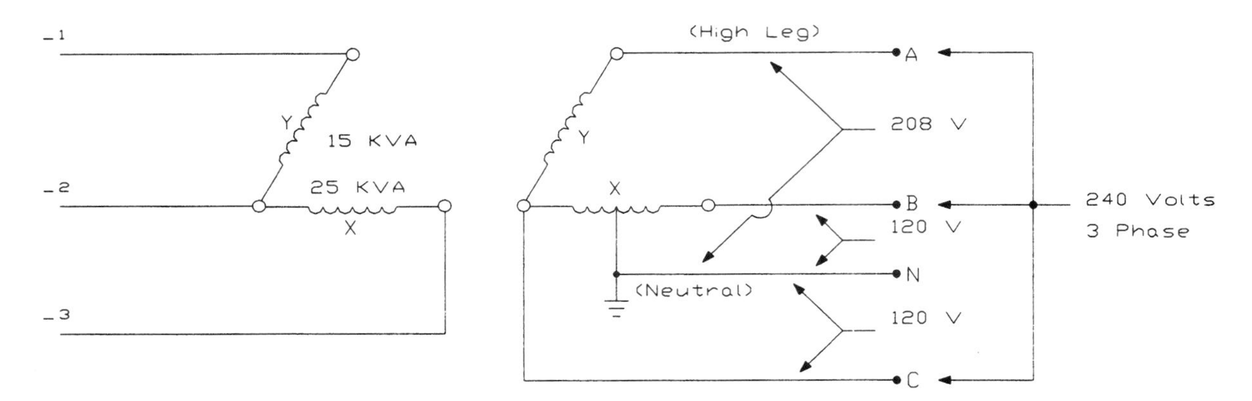

FIGURE 22-2 Open delta connection used to supply both three-phase and single-phase loads

is the high leg. If the voltage is measured from output B to neutral or output C to neutral the voltage will be 120 V. If the voltage is measured from output A to neutral the voltage will be 208 V. The high-leg voltage is 1.732 times higher than either of the other two outputs ($120 \times 1.732 = 208$). The *National Electrical Code*® requires the high leg of a four-wire delta system to be marked by use of an orange wire or by a tag (*NEC*® 215.8).

The voltage and current values of an open delta connection are computed in the same manner as a standard delta-delta connection when three transformers are employed. The voltage and current rules for a delta connection must be used when determining line and phase values of voltage and current.

Name ______________________________ Date ______________

Procedure

1. Using two EMS 8341 transformer modules, connect the circuit shown in Figure 22-3.
2. It will be assumed that a voltage of 208 V will be applied to the primary of the transformer, and the load resistance will be 171.4 Ω per phase.
3. Compute the values listed below.

   $E_{PHASE(pri)}$ = ______________ V

   $E_{PHASE(sec)}$ = ______________ V

   Turns ratio = ______________

   $E_{LINE(sec)}$ = ______________ V

   $E_{PHASE(load)}$ = ______________ V

   $I_{PHASE(load)}$ = ______________ A

   $I_{LINE(sec)}$ = ______________ A

   $I_{PHASE(sec)}$ = ______________ A

   $I_{PHASE(pri)}$ = ______________ A

   $I_{LINE(pri)}$ = ______________ A

4. Set the resistance load module for a resistance of 171.4 Ω per phase.
5. Turn on the power supply and adjust the output voltage for a value of 208 V.

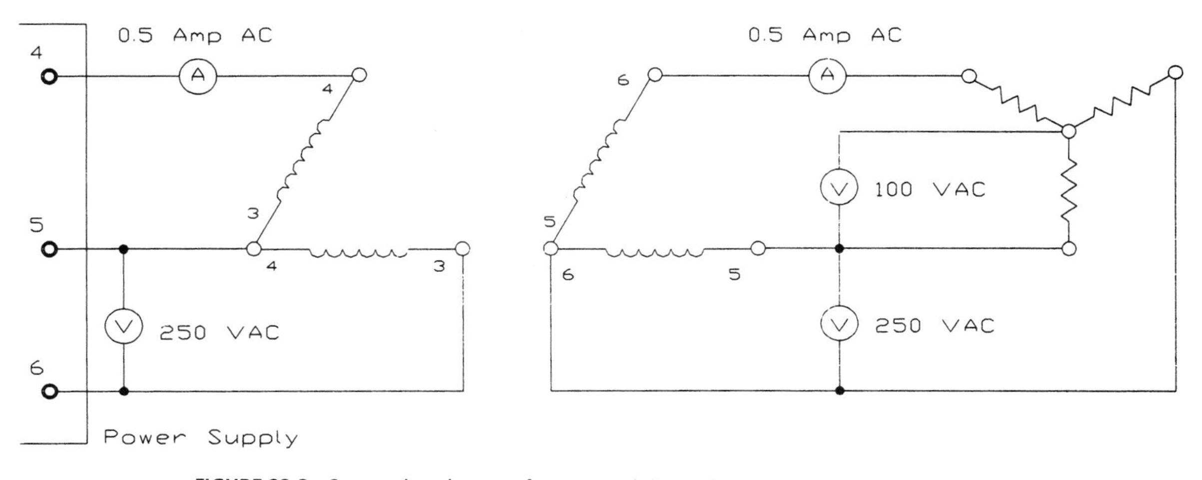

FIGURE 22-3 Connecting the transformer modules to form an open delta connection

6. Measure and record the values listed below.

   $E_{LINE(pri)}$ = ______________ V

   $E_{LINE(sec)}$ = ______________ V

   $E_{PHASE(load)}$ = ______________ V

   $I_{LINE(sec)}$ = ______________ A

   $I_{LINE(pri)}$ = ______________ A

7. Compare the measured values with the computed values.
8. **Return the voltage to 0 V and turn off the power supply.**
9. Disconnect the circuit and return the components to their proper location.

Review Questions

1. How many transformers are needed to make an open delta connection?

   ____

2. Two transformers rated at 100 KVA each are connected in an open delta connection. What is the total output power that can be supplied by this bank?

   ______________ KVA

3. How does the *National Electrical Code*® specify that the high leg of a four-wire delta connection be marked?

   ____

Exercise 23

Single-Phase Loads for Three-Phase Transformers

Objectives

After completing this lab you should be able to:

- Discuss the current and voltage relationships of three-phase transformer connections when connected to single-phase loads.
- Discuss the different types of three-phase transformer connections used to supply single-phase loads.
- Make three-phase transformer connections and measure values of voltage and current for single-phase loads.

Materials and Equipment

Power supply module	EMS 8821
Transformer module (3)	EMS 8341
AC ammeter module	EMS 8425
AC voltmeter module	EMS 8426
Variable-resistance module	EMS 8311
Variable-inductance module	EMS 8321
Separate AC ammeter (0.5 A)	

Discussion

When true three-phase loads are connected to a three-phase transformer bank, there are no problems in balancing the currents and voltages of the individual phases. Figure 23-1 illustrates this condition. In this circuit, a wye-delta three-phase transformer bank is supplying power to a wye-connected three-phase load in which the impedances of the phases are the same. Notice that the amount of current flow in the three phases is the same. This is the ideal condition and is certainly desired for all three-phase transformer loads. Although this is the ideal situation, it is not always possible to obtain a balanced load. Three-phase transformer connections are often used to supply single-phase loads, which

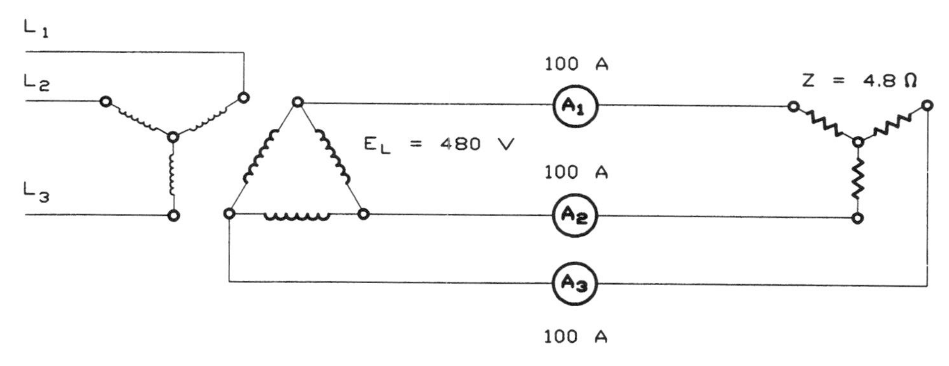

FIGURE 23-1 Wye-delta three-phase transformer bank

tend to unbalance the system. It should be understood that there is no difference between a single-phase load and an unbalanced three-phase load.

OPEN DELTA CONNECTION

The type of three-phase transformer connection used is generally determined by the amount of power needed. When a transformer bank must supply both three-phase and single-phase loads, the utility company will often provide an open delta connection with one transformer center-tapped as shown in Figure 23-2. In this connection, it is assumed that the amount of three-phase power needed is 20 KVA and the amount of single-phase power needed is 30 KVA. Notice that the transformer that has been center-tapped must supply power to both the three-phase and single-phase loads. Because this is an open delta connection, the transformer bank can be loaded to only 86.6% of its full capacity when supplying a three-phase load. The rating of the three-phase transformer bank must therefore be 23.1 KVA in order to supply a load of 20 KVA (20 KVA/0.866 = 23.1 KVA). Because the rating of the two transformers can be added to obtain a total output power rating, one transformer is rated at only half the total amount of power needed, or 12 KVA (23.1 KVA/2 = 11.5 KVA). The transformer that is used to supply power to the three-phase load will be rated at only 12 KVA. The transformer that has been center-tapped must supply power to both the single-phase and three-phase loads. Its capacity will, therefore, be 42 KVA (12 KVA + 30 KVA = 42 KVA).

Voltage Values

The connection shown in Figure 23-2 has a line-to-line voltage of 240 V. The three voltmeters V_1, V_2, and V_3 have all been connected across the three-phase lines and should indicate 240 V each. Voltmeters V_4 and V_5 have been connected between the two lines of the larger transformer and its center tap. These two voltmeters will indicate a voltage of 120 V each. Notice that these two lines and the center tap are used to supply the single-phase power needed. The center tap of the larger transformer is used as a neutral conductor for the single-phase loads. Voltmeter V_6 has been connected between the center tap of the larger transformer and the line of the smaller transformer. This line

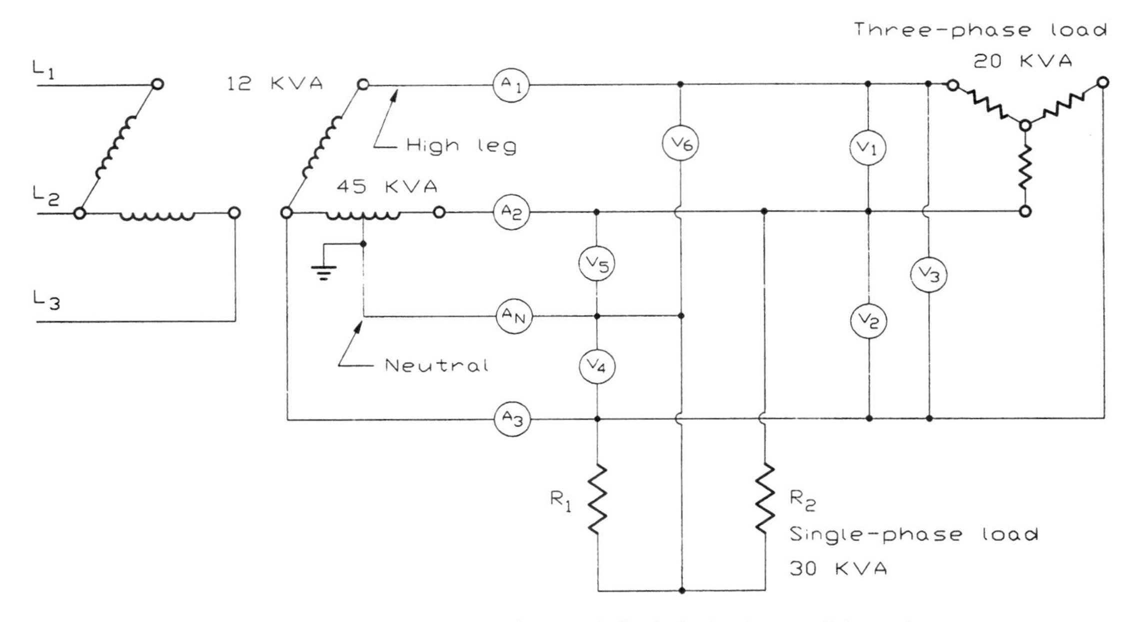

FIGURE 23-2 Open delta connection used to supply both single-phase and three-phase power

is known as a high leg, because the voltage between this line and the neutral conductor will be higher than the voltage between the neutral and either of the other two conductors. The high-leg voltage can be computed by multiplying the single-phase center-tapped voltage value by 1.732. In this case, the high-leg voltage will be 208 V (120 V × 1.732 = 208 V). When this type of connection is employed, the National Electrical Code requires that the high leg be identified by connecting it to an orange wire or by tagging it at any point that it enters an enclosure with the neutral conductor.

Load Conditions

In the first load condition, it will be assumed that only the three-phase load is in operation and none of the single-phase load is operating. If the three-phase load is operating at maximum capacity, ammeters A_1, A_2, and A_3, will indicate a current flow of 48.1 A each (20 KVA/[240 V × 1.732] = 48.1 A). Notice that when only the three-phase load is in operation, the current on each line is balanced.

Now assume that none of the three-phase load is in operation and only the single-phase load is operating. If all the single-phase load is operating at maximum capacity, ammeters A_2 and A_3 will each indicate a value of 125 A (30 KVA/240 V = 125 A). Ammeter A_1 will indicate a current flow of 0 A because all the load is connected between the other two lines of the transformer connection. Ammeter A_N will also indicate a value of 0 A. Ammeter A_N is connected in the neutral conductor, and the neutral conductor carries the sum of the unbalanced load between the two phase conductors. Another way of stating this is to say that the neutral conductor carries the difference between the two line currents. Because these two conductors are now carrying the same amount of current, the difference between them is 0 A.

Now assume that one side of the single-phase load, resistor R_2, has been opened and no current flows through it. If the other line maintains a current flow of 125 A, the neutral conductor will have a current flow of 125 A also (125 A – 0 A = 125 A).

Now assume that resistor R_2 has a value that will permit a current flow of 50 A on that phase. The neutral current will now be 75 A (125 A – 50 A = 75 A). Because the neutral conductor carries the sum of the unbalanced load, the neutral conductor never needs to be larger than the largest line conductor.

It will now be assumed that both three-phase and single-phase loads are operating at the same time. If the three-phase load is operating at maximum capacity and the single-phase load is operating in such a manner that 125 A flows through resistor R_1 and 50 A flows through resistor R_2, the ammeters will indicate the following values:

$$A_1 = 48.1\text{ A}$$

$$A_2 = 98.1\text{ A } (48.1\text{ A} + 50\text{ A} = 98.1\text{ A})$$

$$A_3 = 173.1\text{ A } (125\text{ A} + 48.1\text{ A} = 173.1\text{ A})$$

$$A_N = 75\text{ A } (125\text{ A} - 50\text{ A} = 75\text{ A})$$

Notice that the smaller of the two transformers is supplying current to only the three-phase load, but the larger transformer must supply current for both the single-phase and the three-phase loads.

Although the circuit shown in Figure 23-2 is the most common method of connecting both three-phase and single-phase loads to an open delta transformer bank, it is possible to use the high leg to supply power to a single-phase load also. The circuit shown in Figure 23-3 is a circuit of this type. Resistors R_1 and R_2 are connected to the lines of the transformer that has been center-tapped, and resistor R_3 is connected to the line of the other transformer. If the line-to-line voltage is 240 V, voltmeters V_1 and V_2 will each indicate a value 120 V across resistors R_1 and R_2. Voltmeter V_3, however, will indicate that a voltage of 208 V is applied across resistor R_3.

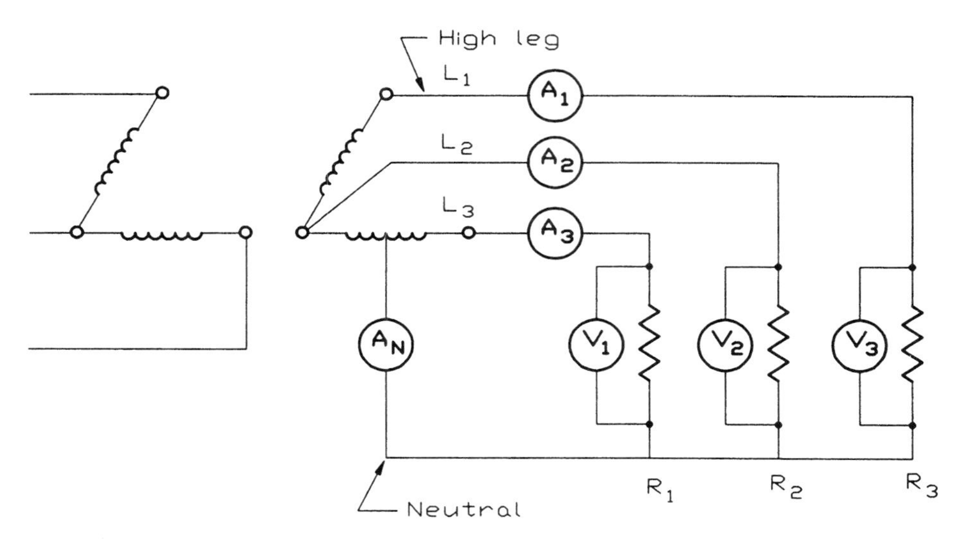

FIGURE 23-3 The high leg supplies power to a single-phase load.

Calculating Neutral Current

The amount of current flow in the neutral conductor will still be the sum of the unbalanced load between lines L_2 and L_3, with the addition of the current flow in the high leg, L_1. To determine the amount of neutral current, use the formula

$$A_N = A_1 + (A_2 - A_3)$$

For example, assume that line L_1 has a current flow of 100 A, line L_2 has a current flow of 75 A, and line L_3 has a current flow of 50 A. The amount of current flow in the neutral conductor would be:

$$A_N = A_1 + (A_2 - A_3)$$
$$A_N = 100\text{ A} + (75\text{ A} - 50\text{ A})$$
$$A_N = 100\text{ A} + 25\text{ A}$$
$$A_N = 125\text{ A}$$

In this circuit, it is possible for the neutral conductor to carry more current than any of the three-phase lines. This circuit is more of an example of why the National Electrical Code requires a high leg to be identified than it is a practical working circuit. The high-leg side of this type of connection will seldom be connected to the neutral conductor.

CLOSED DELTA WITH CENTER TAP

Another three-phase transformer configuration used to supply power to single-phase and three-phase loads is shown in Figure 23-4. This circuit is virtually identical to the circuit shown in Figure 23-2, with the exception that a third transformer has been added to close the delta. Closing the delta permits more power to be supplied for the operation of three-phase loads. In this circuit, it is assumed that the three-phase load has a power requirement of 75 KVA and the single-phase load requires an additional 50 KVA. Three 25-KVA transformers could be used to supply the three-phase power needed (25 KVA × 3 = 75 KVA). The addition of the single-phase load, however, requires one of the transformers to be larger. This transformer must supply both three-phase and single-phase load, which requires it to have a rating of 75 KVA (25 KVA + 50 KVA = 75 KVA).

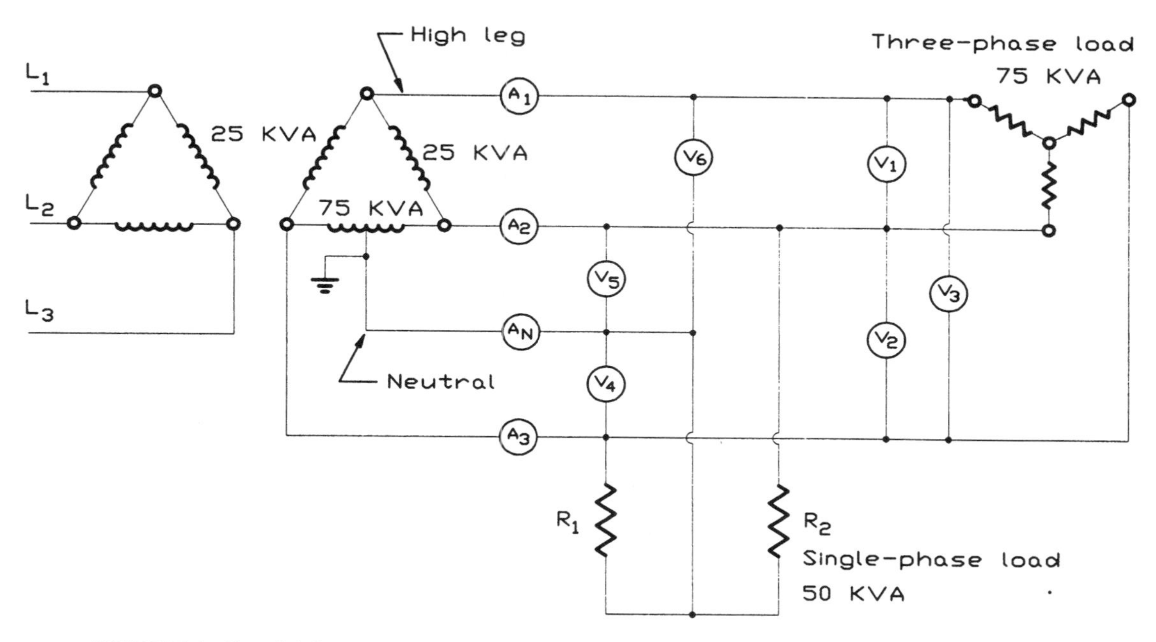

FIGURE 23-4 Closed delta connection used to supply power to both single-phase and three-phase loads

In this circuit, the primary is connected in a delta configuration. Because the secondary side of the transformer bank is a delta connection, either a wye or a delta primary could have been used. This, however, will not be true of all three-phase transformer connections supplying single-phase loads.

CLOSED DELTA WITHOUT CENTER TAP

In the circuit shown in Figure 23-5, the transformer bank has been connected in a wye-delta configuration. Notice that there is no transformer secondary with a center-tapped winding. In this circuit, there is no neutral conductor. The three loads have been connected directly across the three-phase lines. Because these three loads are connected directly across the lines, they form a delta-connected load. If these three loads are intended to be used as single-phase loads, they will, in all likelihood, have changing resistance values. The result of this connection is a three-phase delta-connected load that can be unbalanced in different ways. The amount of current flow in each phase is determined by the impedance of the load and the vectorial relationships of each phase. Each time one of the single-phase loads is altered, the vector relationship changes also. No one phase will become overloaded, however, if the transformer bank has been properly sized for the maximum connected load.

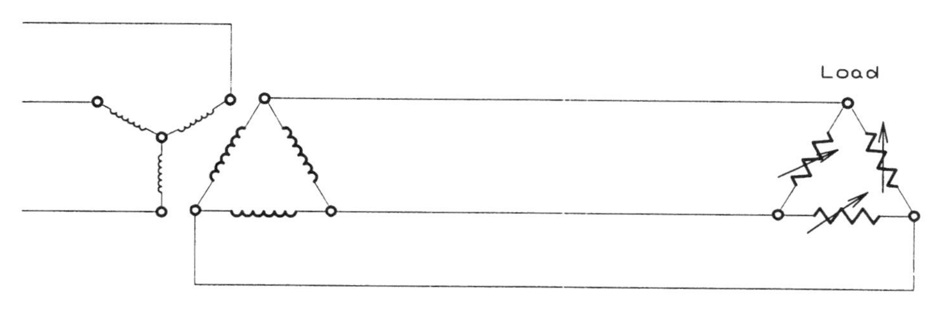

FIGURE 23-5 Three single-phase loads connected as a delta

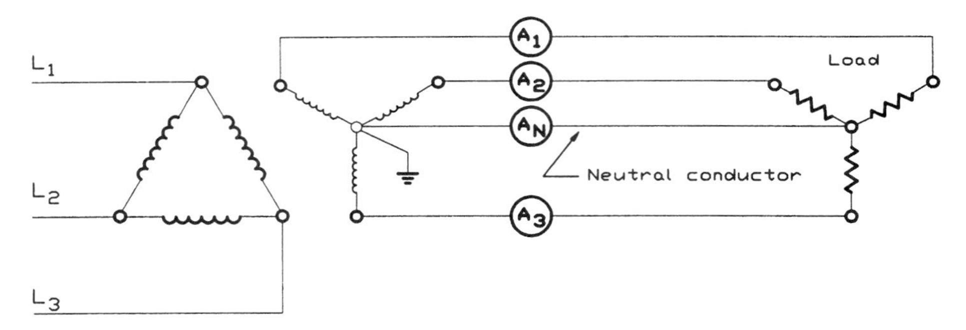

FIGURE 23-6 The center of the wye secondary becomes the neutral conductor.

DELTA-WYE CONNECTION WITH NEUTRAL

The circuit shown in Figure 23-6 is a three-phase transformer connection with a delta-connected primary and wye-connected secondary. The secondary has been center-tapped to form a neutral conductor. This is one of the most common connections used to provide power for single-phase loads. Typical voltages for this type of connection are 208/120 and 480/277. The neutral conductor will carry the sum of the unbalanced current. It should be noted, however, that in this circuit the sum of the unbalanced current is not the difference between two phases. In the delta connection where one transformer was center-tapped to form a neutral conductor, the two lines were 180° out of phase when compared with the center tap. In the wye connection, the lines will be 120° out of phase. When all three lines are carrying the same amount of current, the neutral current will be zero.

It should be noted that a wye-connected secondary with center tap can, under the right conditions, experience extreme unbalance problems.

Caution: ***If this transformer connection is powered by a three-phase three-wire system, the primary winding must be connected in a delta configuration.***

If the primary is connected as a wye connection, the circuit will become exceedingly unbalanced when load is added to the circuit. Connecting the center tap of the primary to the center tap of the secondary will not solve the unbalance problem if a wye primary is used on a three-wire system.

If the incoming power is a three-phase four-wire system as shown in Figure 23-7, however, a wye-connected primary can be used without problem. The neutral conductor connected to the center tap of the primary prevents the unbalance problems. It is a common practice with this type of connection to tie the neutral conductor of both primary and secondary together as shown. It should be noted, however, that when this is done, line isolation between the primary and secondary windings is lost.

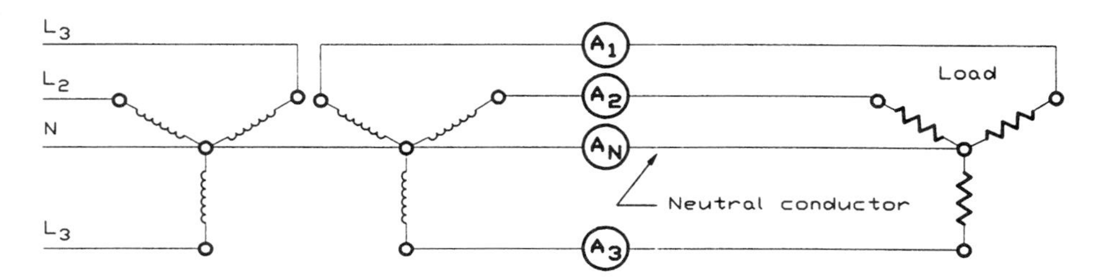

FIGURE 23-7 The center connection of the secondary is connected to the center connection of the primary.

Name ______________________________ Date ______________

Procedure

1. Connect the circuit shown in Figure 23-8.
2. If a line voltage of 208 V is applied to the primary windings of the two transformers, what will be the line voltage of the secondary?

   $E_{S(LINE)}$ = ______________ V

3. Compute the amount of voltage that should be present between the neutral conductor and either of the two line conductors connected to that secondary winding.

$$E_{(LINE\text{-}TO\text{-}NEUTRAL)} = \frac{E_{(LINE)}}{2}$$

   $E_{(LINE\ TO\ NEUTRAL)}$ = ______________ V

4. Compute the amount of voltage that should be present between the neutral conductor and the high-leg conductor.

$$E_{(HIGH\ LEG)} = E_{(LINE\text{-}TO\text{-}NEUTRAL)} \times 1.732$$

   $E_{(HIGH\ LEG)}$ = ______________ V

5. Turn on the power supply and adjust the voltage applied to the primary winding for a value of 208 V.

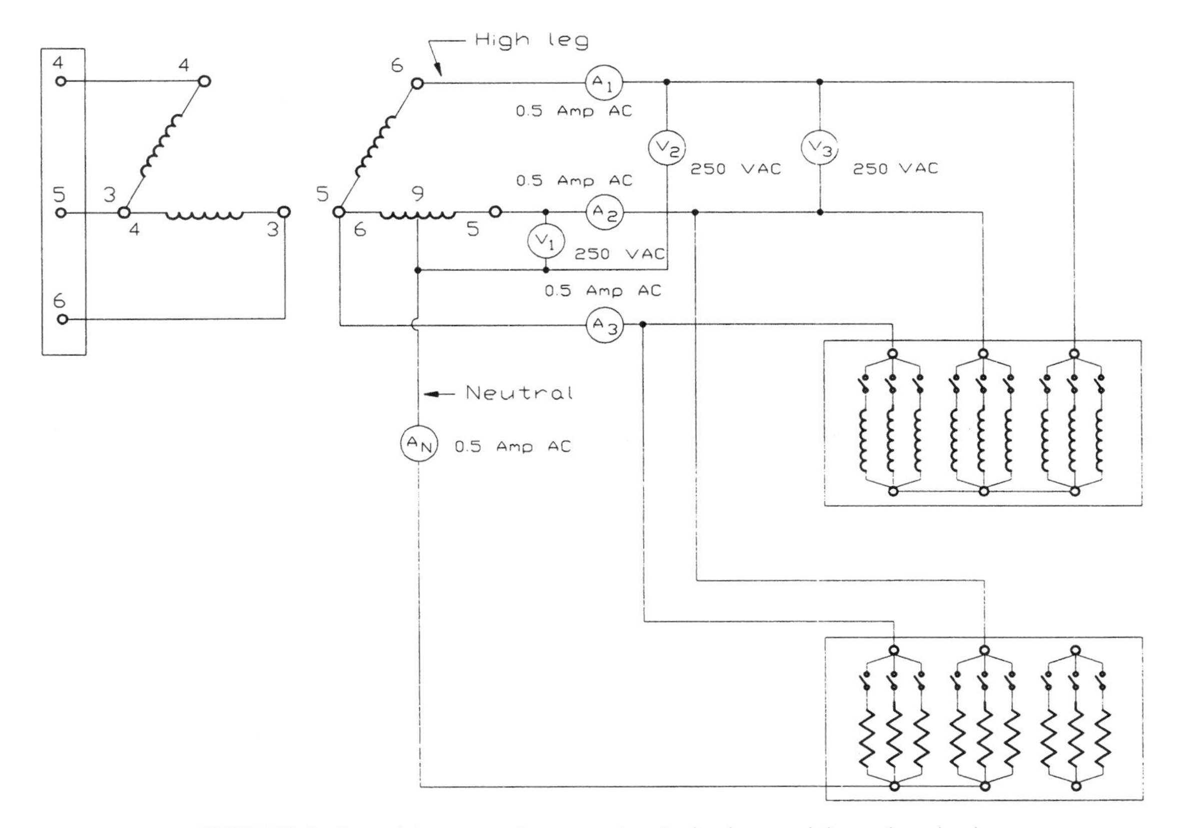

FIGURE 23-8 Open delta connection operating single-phase and three-phase loads

6. Measure the amount of line-to-line voltage using voltmeter V_3, and compare this with the computed value.

   V_3 = ______________ V

7. Measure the amount of line-to-neutral voltage using voltmeter V_1, and compare this value with the computed value.

   V_1 = ______________ V

8. Measure the amount of high-leg voltage using voltmeter V_2, and compare this with the computed value.

   V_2 = ______________ V

9. Adjust the inductive loads to produce an impedance of 600 Ω on each phase. Measure the amount of current on each line and the neutral conductor.

   A_1 = ______________ A

   A_2 = ______________ A

   A_3 = ______________ A

   A_N = ______________ A

10. Are the current values indicated by ammeters A1, A2, and A3 approximately equal in value?

    ____

11. Is there any current flow indicated by ammeter A_N?

    ____

12. Adjust the resistive load bank so that the first section of resistors has a value of 300 Ω.

13. List the voltages and currents indicated by the voltmeters and ammeters.

    A_1 = ______________ A V_1 = ______________ V

    A_2 = ______________ A V_2 = ______________ V

    A_3 = ______________ A V_3 = ______________ V

    A_N = ______________ A

14. When load was added to the resistive load bank, which simulates the addition of a single-phase load, did the voltage values change from their previous values?

    ____

15. Adjust the resistive load bank so that the first bank of resistors has a total resistance value of 200 Ω.

16. List the voltages and currents indicated by the voltmeters and ammeters.

    A_1 = ______________ A V_1 = ______________ V

    A_2 = ______________ A V_2 = ______________ V

    A_3 = ______________ A V_3 = ______________ V

    A_N = ______________ A

17. Did the amount of neutral current increase or decrease in value?

    ____

18. Did the voltage values change or remain approximately the same?

____

19. Adjust the resistive load bank so that the second bank of resistors has a resistance value of 400 Ω. Do not change the setting of the first bank of resistors.

20. List the voltages and currents indicated by the voltmeters and ammeters.

A_1 = _______________ A V_1 = _______________ V

A_2 = _______________ A V_2 = _______________ V

A_3 = _______________ A V_3 = _______________ V

A_N = _______________ A

21. Did the neutral current increase or decrease in value?

____

22. Did the voltage values change or remain approximately the same?

____

23. Adjust the resistive load bank so that the second bank of resistors has a resistance value of 300 Ω. Do not change the resistance value of the first bank of resistors.

24. List the voltages and currents indicated by the voltmeters and ammeters.

A_1 = _______________ A V_1 = _______________ V

A_2 = _______________ A V_2 = _______________ V

A_3 = _______________ A V_3 = _______________ V

A_N = _______________ A

25. Did the neutral current increase or decrease in value?

____

26. Did the voltage values change or remain approximately the same?

____

27. Adjust the resistance load bank so that the second bank of resistors has a value of 200 Ω. Do not change the resistance value of the first bank of resistors.

28. List the voltages and currents indicated by the voltmeters and ammeters.

A_1 = _______________ A V_1 = _______________ V

A_2 = _______________ A V_2 = _______________ V

A_3 = _______________ A V_3 = _______________ V

A_N = _______________ A

29. The neutral current value should now be zero. Both single-phase loads are now the same, causing the amount of current flow through them to be the same. Because the neutral current is the sum of the unbalanced load, or the difference between the two currents, it should be zero. Is the neutral current zero at this point?

____

30. **Return the primary voltage to 0 V and turn off the power supply. Open all switches on both the inductive and resistive load banks.**
31. Connect the circuit shown in Figure 23-9.
32. Turn on the power supply and adjust the voltage for a value of 208 V.
33. Refer to the voltage rating shown on the transformer module. What is the line-to-line voltage of the secondary?

    ____________ V

34. Measure the voltages indicated by voltmeters V_1, V_2, and V_3.

    V_1 = ____________ V V_2 = ____________ V V_3 = ____________ V

35. Do the measured voltage values agree with the values listed on the transformer modules?

    ____

36. Adjust the resistance load bank so that the first section of resistors has a resistance of 400 Ω.
37. List the voltages and currents indicated by the voltmeters and ammeters.

    A_1 = ____________ A V_1 = ____________ V
    A_2 = ____________ A V_2 = ____________ V
    A_3 = ____________ A V_3 = ____________ V

38. Did the voltage values change or remain approximately the same?

    ____

39. Adjust the resistance load bank so that the second bank of resistors has a resistance of 300 Ω.

    NOTE: It may be necessary to change the ammeter range from 0.5 A to 2.5 A as loads are changed in the remainder of this experiment to prevent overloading the ammeters. For greater accuracy in measurement, however, ammeters should be set on the lowest possible range that can be used without pegging the meter.

40. List the voltages and currents indicated by the voltmeters and ammeters.

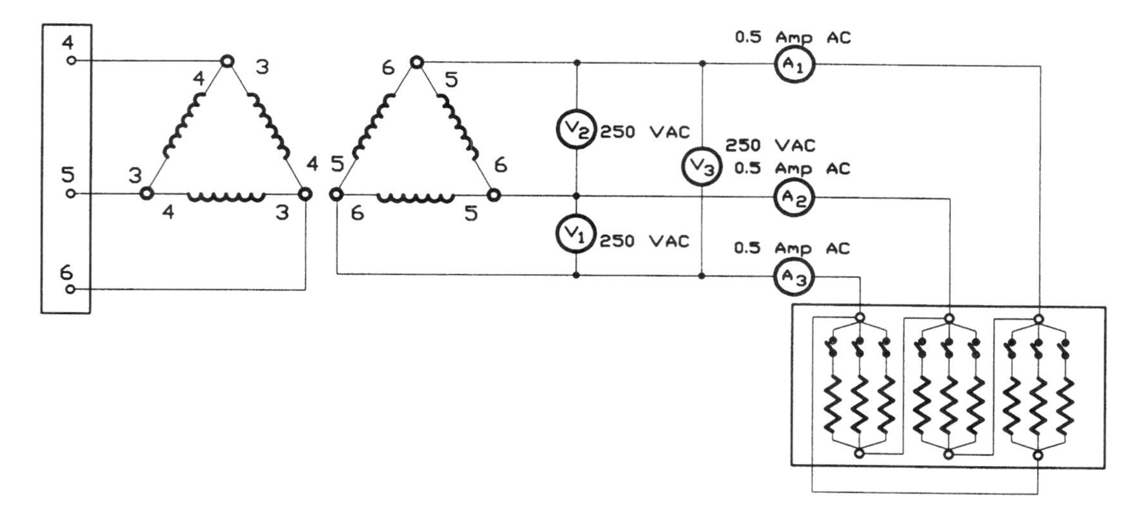

FIGURE 23-9 Delta-delta transformer connected to a delta load

A_1 = ______________ A V_1 = ______________ V

A_2 = ______________ A V_2 = ______________ V

A_3 = ______________ A V_3 = ______________ V

41. Did the voltage values change or remain approximately the same?

____

42. Adjust the resistance load bank so that the third section of resistors has a resistance value of 240 Ω.

43. List the voltages and currents indicated by the voltmeters and ammeters.

A_1 = ______________ A V_1 = ______________ V

A_2 = ______________ A V_2 = ______________ V

A_3 = ______________ A V_3 = ______________ V

44. Did the voltage values change or remain approximately the same?

____

45. Adjust the resistance load bank so that each section has a resistance of 240 Ω.

46. List the voltages and currents indicated by the voltmeters and ammeters.

A_1 = ______________ A V_1 = ______________ V

A_2 = ______________ A V_2 = ______________ V

A_3 = ______________ A V_3 = ______________ V

47. Did the voltage values change or remain approximately the same?

____

48. **Return the power supply voltage to 0 V and turn off the power supply. Open all the switches on the resistance load bank.**

49. Connect the circuit shown in Figure 23-10.

50. Turn on the power supply and adjust the voltage for a value of 208 V.

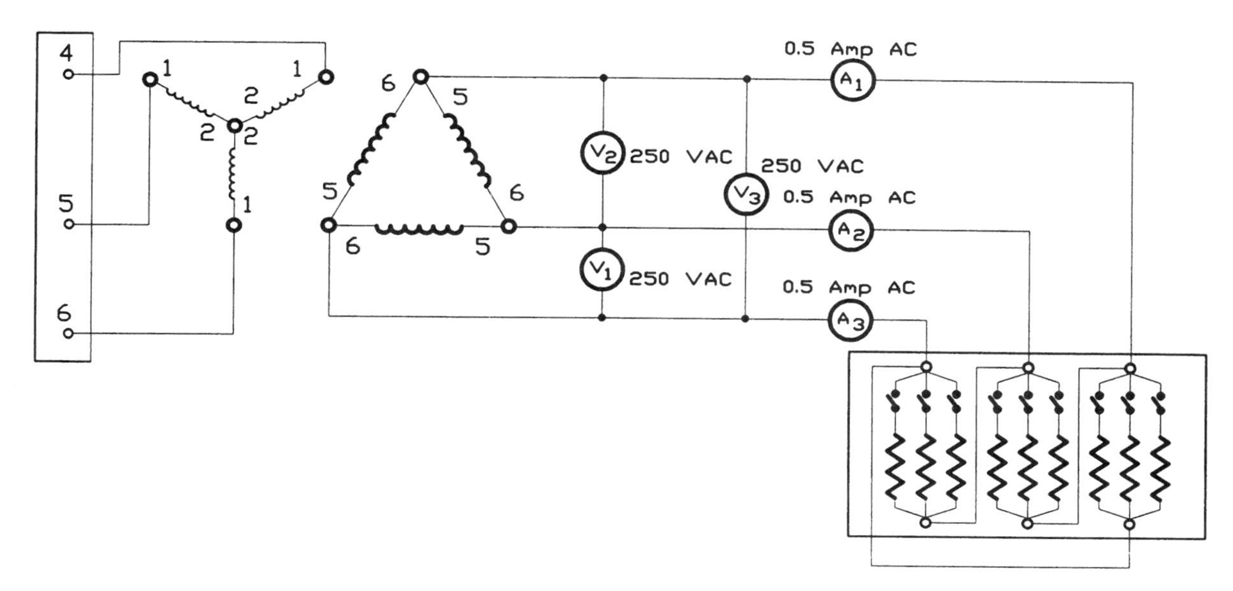

FIGURE 23-10 Wye-delta three-phase transformer connection

51. Refer to the voltage rating shown on the transformer module. What is the line-to-line voltage of the secondary?

    ______________ V

52. Measure the voltages indicated by voltmeters V_1, V_2, and V_3.

    V_1 = ______________ V V_2 = ______________ V V_3 = ______________ V

53. Do the measured voltage values agree with the values listed on the transformer modules?

    ____

54. Adjust the resistance load bank so that the first section of resistors has a resistance of 400 Ω.

55. List the voltages and currents indicated by the voltmeters and ammeters.

    A_1 = ______________ A V_1 = ______________ V

    A_2 = ______________ A V_2 = ______________ V

    A_3 = ______________ A V_3 = ______________ V

56. Did the voltage values change or remain approximately the same?

    ____

57. Adjust the resistance load bank so that the second bank of resistors has a resistance of 300 Ω.

58. List the voltages and currents indicated by the voltmeters and ammeters.

    A_1 = ______________ A V_1 = ______________ V

    A_2 = ______________ A V_2 = ______________ V

    A_3 = ______________ A V_3 = ______________ V

59. Did the voltage values change or remain approximately the same?

    ____

60. Adjust the resistance load bank so that the third section of resistors has a resistance value of 240 Ω.

61. List the voltages and currents indicated by the voltmeters and ammeters.

    A_1 = ______________ A V_1 = ______________ V

    A_2 = ______________ A V_2 = ______________ V

    A_3 = ______________ A V_3 = ______________ V

62. Did the voltage values change or remain approximately the same?

    ____

63. Adjust the resistance load bank so that each section has a resistance of 240 Ω.

64. List the voltages and currents indicated by the voltmeters and ammeters.

    A_1 = ______________ A V_1 = ______________ V

    A_2 = ______________ A V_2 = ______________ V

    A_3 = ______________ A V_3 = ______________ V

65. Did the voltage values change or remain approximately the same?

____

66. **Return the power supply voltage to 0 V and turn off the power supply. Open all the switches on the resistance load bank.**

67. Is there any noticeable difference, as far as balanced voltages are concerned, between the primary of the transformer bank being connected in a delta configuration and its being connected in a wye configuration?

____

68. Connect the circuit shown in Figure 23-11.

69. Assuming that 208 V is to be connected to the primary winding of the transformer, what should be the phase voltage of the secondary? This value is listed on the transformer module.

    $E_{(PHASE)}$ = _______________ V

70. The phase value of voltage of the secondary winding will be the voltage applied across each section of the load resistors.

71. Turn on the power supply and adjust the output voltage for a value of 208 V.

72. What values of voltage and current are indicated by the ammeters and voltmeters?

    A_1 = _______________ A V_1 = _______________ V

    A_2 = _______________ A V_2 = _______________ V

    A_3 = _______________ A V_3 = _______________ V

    A_N = _______________ A

73. Do the voltage readings correspond with the listed values?

____

74. Set the first section of the resistance bank for a value of 300 Ω.

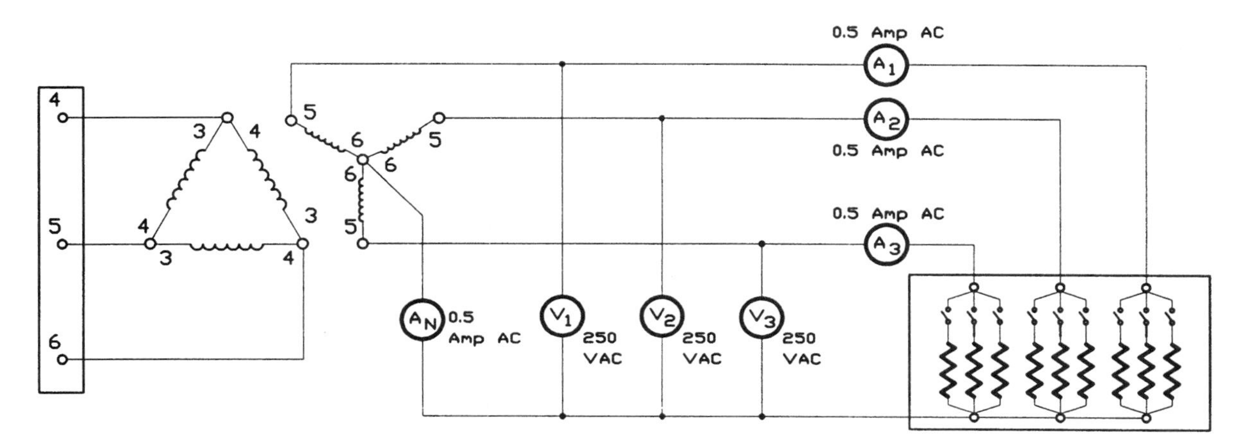

FIGURE 23-11 Delta-wye three-phase transformer connection

75. What values of voltage and current are indicated by the ammeters and voltmeters?

A_1 = ______________ A V_1 = ______________ V

A_2 = ______________ A V_2 = ______________ V

A_3 = ______________ A V_3 = ______________ V

A_N = ______________ A

76. Set the second section of the resistor bank for a value of 400 Ω.

77. What values of voltage and current are indicated by the ammeters and voltmeters?

A_1 = ______________ A V_1 = ______________ V

A_2 = ______________ A V_2 = ______________ V

A_3 = ______________ A V_3 = ______________ V

A_N = ______________ A

78. Set the third section of resistors for a value of 240 Ω.

79. What values of voltage and current are indicated by the ammeters and voltmeters?

A_1 = ______________ A V_1 = ______________ V

A_2 = ______________ A V_2 = ______________ V

A_3 = ______________ A V_3 = ______________ V

A_N = ______________ A

80. Set all three sections of the resistor bank for a value of 240 Ω each.

81. What values of voltage and current are indicated by the ammeters and voltmeters?

A_1 = ______________ A V_1 = ______________ V

A_2 = ______________ A V_2 = ______________ V

A_3 = ______________ A V_3 = ______________ V

A_N = ______________ A

82. Did the voltages become unbalanced as load was added to the circuit?

____

83. Did the neutral current ever become higher than the highest current flow in a single phase?

____

84. **Return the voltage to 0 V and turn off the power supply. Open all switches on the resistance load bank.**

85. Connect the circuit shown in Figure 23-12.

86. Assuming that 208 V is to be connected to the primary winding of the transformer, what should be the phase voltage of the primary? What should be phase value of voltage of the secondary? This value is listed on the transformer module.

$E_{PHASE(pri)}$ = ______________ V

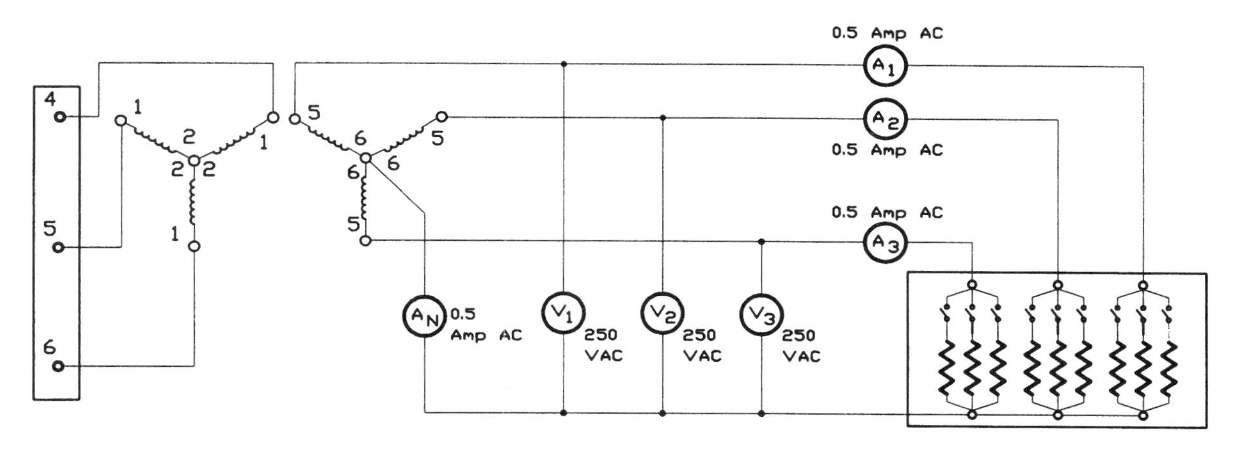

FIGURE 23-12 Wye-wye three-phase transformer connection

87. The phase value of voltage of the secondary winding will be the voltage applied across each section of the load resistors.
88. Turn on the power supply and adjust the output voltage for a value of 208 V.
89. What values of voltage and current are indicated by the ammeters and voltmeters?

A_1 = ______________ A V_1 = ______________ V

A_2 = ______________ A V_2 = ______________ V

A_3 = ______________ A V_3 = ______________ V

A_N = ______________ A

90. Do the voltage readings correspond with the listed values?

____

91. Set the first section of the resistance bank for a value of 300 Ω.
92. What values of voltage and current are indicated by the ammeters and voltmeters?

A_1 = ______________ A V_1 = ______________ V

A_2 = ______________ A V_2 = ______________ V

A_3 = ______________ A V_3 = ______________ V

A_N = ______________ A

93. Set the second section of the resistor bank for a value of 400 Ω.
94. What values of voltage and current are indicated by the ammeters and voltmeters?

A_1 = ______________ A V_1 = ______________ V

A_2 = ______________ A V_2 = ______________ V

A_3 = ______________ A V_3 = ______________ V

A_N = ______________ A

95. Set the third section of resistors for a value of 240 Ω.

96. What values of voltage and current are indicated by the ammeters and voltmeters?

A_1 = ______________ A V_1 = ______________ V

A_2 = ______________ A V_2 = ______________ V

A_3 = ______________ A V_3 = ______________ V

A_N = ______________ A

97. Set all three sections of the resistor bank for a value of 240 Ω each.

98. What values of voltage and current are indicated by the ammeters and voltmeters?

A_1 = ______________ A V_1 = ______________ V

A_2 = ______________ A V_2 = ______________ V

A_3 = ______________ A V_3 = ______________ V

A_N = ______________ A

99. Did the voltages become unbalanced as load was added to the circuit?

____

100. **Return the voltage to 0 V and turn off the power supply. Open all switches on the resistance load bank.**

101. Connect the circuit shown in Figure 23-13. Notice that this is the same as the circuit shown in Figure 23-12 except that the center tap of the primary winding has been connected to the neutral of the incoming power.

102. Turn on the power supply and adjust the output voltage for a value of 208 V.

103. What values of voltage and current are indicated by the ammeters and voltmeters?

A_1 = ______________ A V_1 = ______________ V

A_2 = ______________ A V_2 = ______________ V

A_3 = ______________ A V_3 = ______________ V

A_N = ______________ A

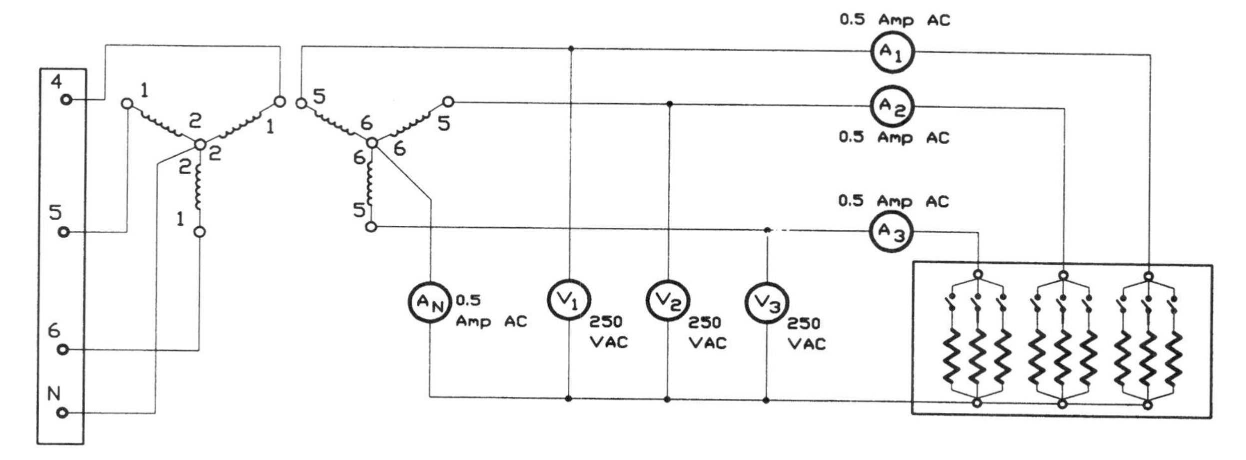

FIGURE 23-13 The center point of the primary is connected to neutral

104. Do the voltage readings correspond with the listed values?

105. Set the first section of the resistance bank for a value of 300 Ω

106. What values of voltage and current are indicated by the ammeters and voltmeters?

A_1 = ______________ A V_1 = ______________ V

A_2 = ______________ A V_2 = ______________ V

A_3 = ______________ A V_3 = ______________ V

A_N = ______________ A

107. Set the second section of the resistor bank for a value of 400 Ω.

108. What values of voltage and current are indicated by the ammeters and voltmeters?

A_1 = ______________ A V_1 = ______________ V

A_2 = ______________ A V_2 = ______________ V

A_3 = ______________ A V_3 = ______________ V

A_N = ______________ A

109. Set the third section of resistors for a value of 240 Ω.

110. What values of voltage and current are indicated by the ammeters and voltmeters?

A_1 = ______________ A V_1 = ______________ V

A_2 = ______________ A V_2 = ______________ V

A_3 = ______________ A V_3 = ______________ V

A_N = ______________ A

111. Set all three sections of the resistor bank for a value of 240 Ω each.

112. What values of voltage and current are indicated by the ammeters and voltmeters?

A_1 = ______________ A V_1 = ______________ V

A_2 = ______________ A V_2 = ______________ V

A_3 = ______________ A V_3 = ______________ V

A_N = ______________ A

113. Did the voltages become unbalanced as load was added to the circuit?

____

114. Did the neutral current ever become higher than the highest current flow in a single phase?

____

115. **Return the voltage to 0 V and turn off the power supply. Open all switches on the resistance load bank.**

116. Connect the circuit shown in Figure 23-14. Notice that this is the same circuit except that a connection has been added between the center tap of the primary and secondary windings of the transformer bank.

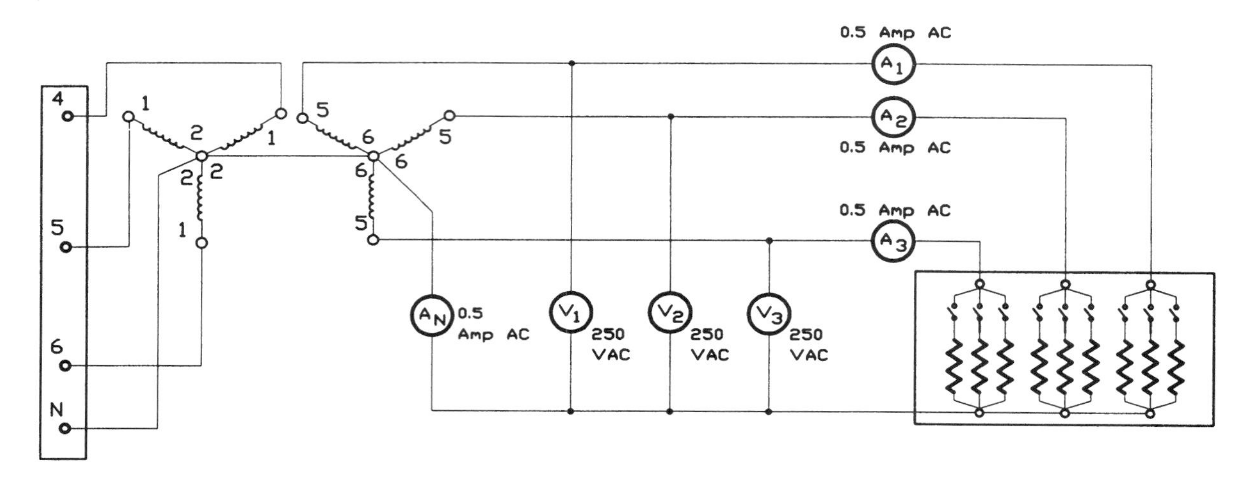

FIGURE 23-14 The center points of the primary and secondary are connected together.

117. Turn on the power supply and adjust the output voltage for a value of 208 V.

118. What values of voltage and current are indicated by the ammeters and voltmeters?

A_1 = ______________ A	V_1 = ______________ V
A_2 = ______________ A	V_2 = ______________ V
A_3 = ______________ A	V_3 = ______________ V
A_N = ______________ A	

119. Do the voltage readings correspond with the listed values?

____

120. Set the first section of the resistance bank for a value of 300 Ω.

121. What values of voltage and current are indicated by the ammeters and voltmeters?

A_1 = ______________ A	V_1 = ______________ V
A_2 = ______________ A	V_2 = ______________ V
A_3 = ______________ A	V_3 = ______________ V
A_N = ______________ A	

122. Set the second section of the resistor bank for a value of 400 Ω.

123. What values of voltage and current are indicated by the ammeters and voltmeters?

A_1 = ______________ A	V_1 = ______________ V
A_2 = ______________ A	V_2 = ______________ V
A_3 = ______________ A	V_3 = ______________ V
A_N = ______________ A	

124. Set the third section of resistors for a value of 240 Ω.

125. What values of voltage and current are indicated by the ammeters and voltmeters?

A_1 = ______________ A V_1 = ______________ V

A_2 = ______________ A V_2 = ______________ V

A_3 = ______________ A V_3 = ______________ V

A_N = ______________ A

126. Set all three sections of the resistor bank for a value of 240 Ω each.

127. What values of voltage and current are indicated by the ammeters and voltmeters?

A_1 = ______________ A V_1 = ______________ V

A_2 = ______________ A V_2 = ______________ V

A_3 = ______________ A V_3 = ______________ V

A_N = ______________ A

128. Did the voltages become unbalanced as load was added to the circuit?

____

129. Did the neutral current ever become higher than the highest current flow in a single phase?

____

130. **Return the voltage to 0 V and turn off the power supply. Open all switches on the resistance load bank.**

131. Connect the circuit shown in Figure 23-15. Notice that this is the same circuit except that the connection has been removed from the center tap of the primary to the neutral of the incoming power supply. The center taps of the secondary windings are still connected to the center tap of the primary windings, however. It will be seen that connecting the center taps of the primary and secondary windings together will not solve the voltage unbalance problems when load is added to the secondary. In a wye-connected primary and a wye-connected secondary with single-phase loads, the incoming power must be a three-phase four-wire system.

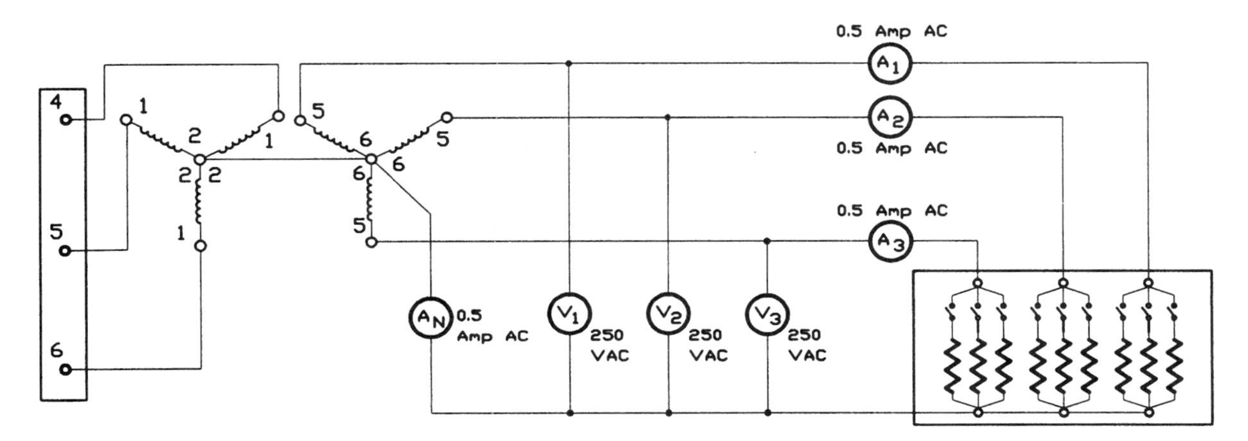

FIGURE 23-15 Wye-wye connection operated without a neutral conductor connected to the primary

132. Turn on the power supply and adjust the output voltage for a value of 208 V.

133. What values of voltage and current are indicated by the ammeters and voltmeters?

A_1 = ______________ A V_1 = ______________ V

A_2 = ______________ A V_2 = ______________ V

A_3 = ______________ A V_3 = ______________ V

A_N = ______________ A

134. Do the voltage readings correspond with the listed values?

____

135. Set the first section of the resistance bank for a value of 300 Ω.

136. What values of voltage and current are indicated by the ammeters and voltmeters?

A_1 = ______________ A V_1 = ______________ V

A_2 = ______________ A V_2 = ______________ V

A_3 = ______________ A V_3 = ______________ V

A_N = ______________ A

137. Set the second section of the resistor bank for a value of 400 Ω.

138. What values of voltage and current are indicated by the ammeters and voltmeters?

A_1 = ______________ A V_1 = ______________ V

A_2 = ______________ A V_2 = ______________ V

A_3 = ______________ A V_3 = ______________ V

A_N = ______________ A

139. Set the third section of resistors for a value of 240 Ω.

140. What values of voltage and current are indicated by the ammeters and voltmeters?

A_1 = ______________ A V_1 = ______________ V

A_2 = ______________ A V_2 = ______________ V

A_3 = ______________ A V_3 = ______________ V

A_N = ______________ A

141. Set all three sections of the resistor bank for a value of 240 Ω each.

142. What values of voltage and current are indicated by the ammeters and voltmeters?

A_1 = ______________ A V_1 = ______________ V

A_2 = ______________ A V_2 = ______________ V

A_3 = ______________ A V_3 = ______________ V

A_N = ______________ A

143. Did the voltages become unbalanced as load was added to the circuit?

____

144. **Return the voltage to 0 V and turn off the power supply. Disconnect the circuit and return the components to their proper place.**

Review Questions

1. An open delta three-phase transformer system has one transformer center-tapped to provide a neutral for single-phase voltages. If the voltage from line to center tap is 277 V, what is the high-leg voltage?

   ______________ V

2. If a single-phase load is connected across the two line conductors and neutral of the above transformer, and one line has a current of 80 A and the other line has a current of 68 A, how much current is flowing in the neutral conductor?

   ______________ A

3. A three-phase transformer connection has a delta-connected secondary, and one of the transformers has been center-tapped to form a neutral conductor. The phase-to-neutral value of the center-tapped secondary winding is 120 V. If the high leg is connected to a single-phase load, how much voltage will be applied to that load?

   ______________ V

4. A three-phase transformer connection has a delta-connected primary and a wye-connected secondary. The center tap of the wye is used as a neutral conductor. If the line-to-line voltage is 480 V, what is the voltage between any one phase conductor and the neutral conductor?

   ______________ V

5. A three-phase transformer bank has the secondary connected in a wye configuration. The center tap is used as a neutral conductor. If the voltage across any phase conductor and neutral is 120 V, how much voltage would be applied to a three-phase load connected to the secondary of this transformer bank?

   ______________ V

6. A three-phase transformer bank has the primary and secondary windings connected in a wye configuration. The secondary center tap is being used as a neutral to supply single-phase loads. Will connecting the center tap connection of the secondary to the center tap connection of the primary permit the secondary voltage to stay in balance when a single-phase load is added to the secondary?

   ____

7. Refer to the transformer connection in question 6. If the center tap of the primary is connected to a neutral conductor on the incoming power, will it permit the secondary voltages to be balanced when single-phase loads are added?

   ____

Exercise 24

Direct Current Machine Characteristics

Objectives

After completing this lab you should be able to:

- Make a visual inspection of a direct current machine.
- Identify the armature, series field winding, and shunt field winding, with an ohmmeter.
- Measure the resistance of the armature, series field winding, and shunt field winding.

Materials and Equipment

Power supply module	EMS 8821
Direct current machine module	EMS 8211
Electrodynamometer module	EMS 8911
or prime mover/dynamometer module	EMS 8960
DC metering module	EMS 8412
Ohmmeter (supplied by student)	

Discussion

The direct current machine used in the EMS training system is designed to be used as both a generator and a motor. This is true for almost any direct current machine. The type of armature winding and field winding used in DC machines is determined by the application of the machine. The starter motor on an automobile, for example, must have extremely high torque and operate on a 12-V source. Starter motors are series motors and, therefore, do not contain a shunt field winding. Large-horsepower DC generators and motors found in industry are generally of the compound type and contain both series and shunt field windings. These machines will commonly operate on voltages that range from 125 to 1200 VDC.

THE ARMATURE

The armature is the rotating member of a direct current machine. It is made on a laminated iron core. Slots are cut in the core to permit windings to be inserted, as shown in Figure 24-1. It is common practice to insert more than one set of windings in a slot.

The armature also contains the commutator. The commutator is made of copper segments separated by insulating material. The windings of the armature are connected to the commutator bars. The commutator performs two basic functions. One function is to connect the armature windings to the external circuit via the brushes. The second function is to maintain the proper direction of current flow through the armature. When the DC machine is used as a generator, the commutator operates as a mechanical rectifier and converts the AC voltage produced in the armature into DC voltage before it is applied to the external circuit.

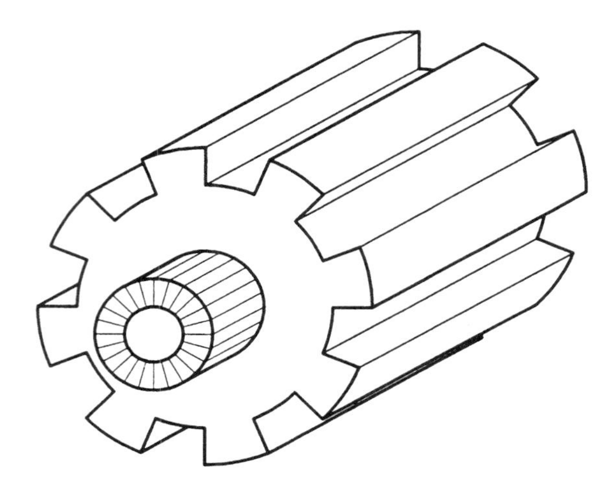

FIGURE 24-1 Armature slots and commutator bars

THE BRUSHES

The brushes are generally made from a carbon material that is softer than the material the commutator is made of. This is to promote wear of the brushes instead of the commutator. The function of the brushes is to provide connection to the commutator. The brush leads of a DC machine are generally marked A_1 and A_2, and are referred to as the armature leads. The armature leads are marked 1 and 2 on the Lab-Volt EMS 8211 direct current machine module, however.

THE SERIES FIELD

The series field of a DC machine is made with a few turns of relatively large wire (relative to the shunt field winding). The series and shunt field windings are wound around the same pole pieces. The series field winding, however, is intended to be connected in series with the armature and must, therefore, have a low resistance. The series field winding of most DC machines found in industry is marked with an S_1 and S_2. The series field winding is marked 3 and 4 on the EMS 8211 direct current machine module.

THE SHUNT FIELD

The shunt field is made with many turns of small wire. It is intended to be connected in parallel with the armature. Because the shunt field winding is connected in parallel with the armature, it will have the same voltage applied to it that the armature has. The amount of current flow through the shunt field winding is limited by the resistance of the winding. For this reason, the shunt field winding will have a much higher resistance than the series field winding. The shunt field leads of most DC machines are marked F_1 and F_2 and are generally referred to as the field leads. The shunt field leads are marked 5 and 6 on the EMS 8211 direct current machine module.

THE SHUNT FIELD RHEOSTAT

The EMS 8211 direct current machine module contains a shunt field rheostat. This rheostat is designed to be connected in series with the shunt field and to control current flow through the field by inserting or removing resistance. The rheostat connection is marked with terminal numbers 7 and 8.

Name ______________________________ Date ______________

Procedure

1. Remove the EMS 8211 direct current machine module, Figure 24-2 from the mobile console.
2. Examine the direct current machine and locate the following items:
   A. The brushes, located on the commutator.

   B. The brush position adjustment control, used to adjust the position of the brushes on the commutator. If you face the motor from the front, the adjustment handle will be located at the top of the motor. A small red dot on the motor frame marks the approximated position of the neutral plane. Move the control handle back and forth and observe that the brushes change position on the commutator. Return the brushes to their approximate neutral position.

   C. The series field winding, wound around each pole piece. Two separate windings are wound around each pole piece. The winding with the larger wire and fewer turns is the series field.

   D. The shunt field winding, the winding made with the smaller wire and more turns.

   E. The armature, the rotating member of the machine. Notice that the armature is constructed with coils of wire laid into slots cut into the armature core. The coils of wire are connected to the segments of the commutator.

   F. The shunt field circuit breaker, located in the lower front of the module. The circuit breaker can be seen by looking under the bottom of the module. If you trace the wire connections, you can see that the circuit breaker is connected in series with one of the shunt field rheostat leads. When this machine is operated, the shunt field should be connected in series with the shunt field rheostat, or the circuit breaker cannot protect the shunt field winding.

   G. The shunt field rheostat, located on the front panel of the module.
3. Mark one of the commutator bars with a piece of chalk or other material that is easily removed. Count the number of commutator bars.

____

FIGURE 24-2 Direct current machine module (Courtesy of Lab-Volt® Systems, Inc.)

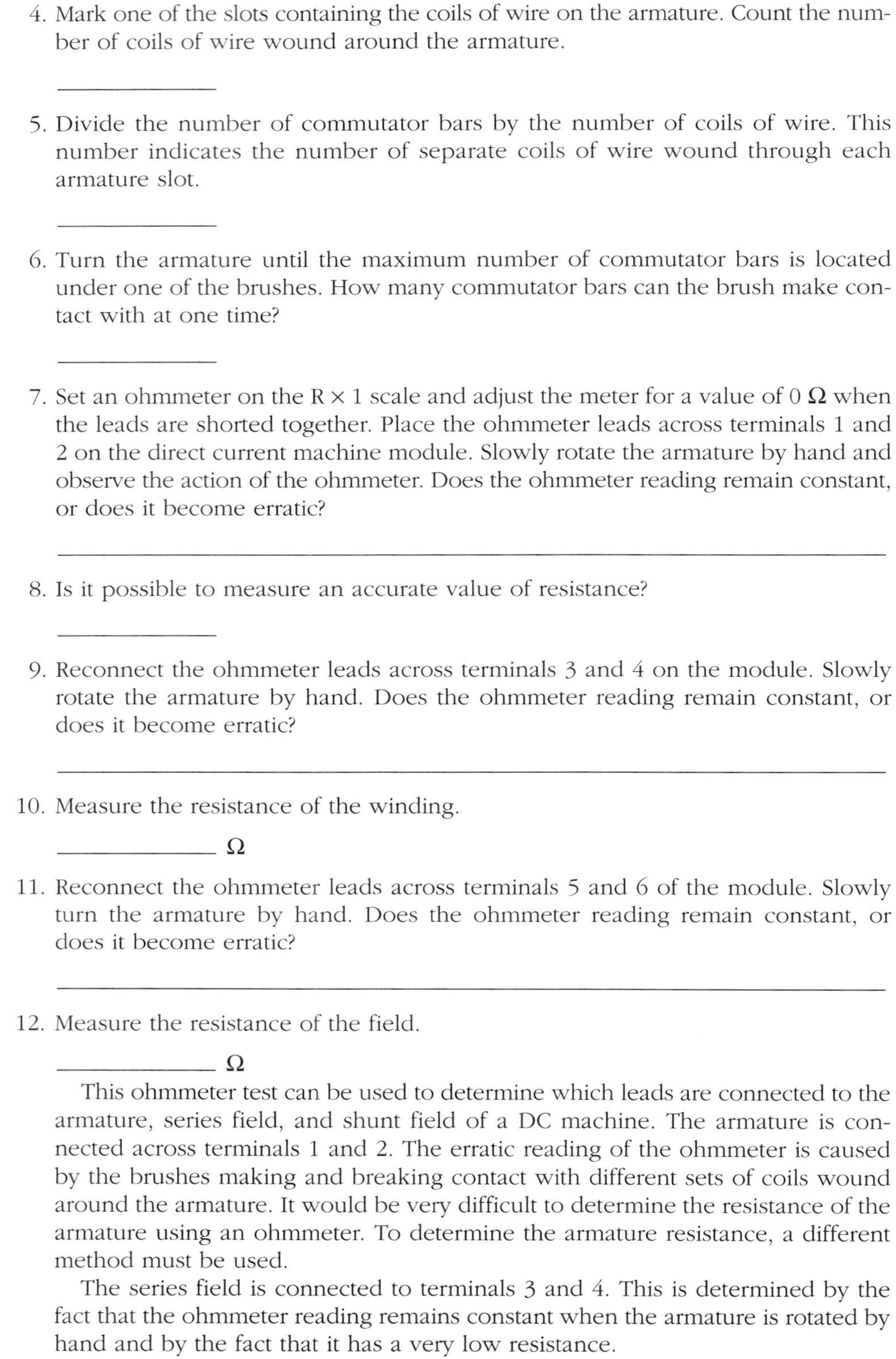

4. Mark one of the slots containing the coils of wire on the armature. Count the number of coils of wire wound around the armature.

   ____

5. Divide the number of commutator bars by the number of coils of wire. This number indicates the number of separate coils of wire wound through each armature slot.

   ____

6. Turn the armature until the maximum number of commutator bars is located under one of the brushes. How many commutator bars can the brush make contact with at one time?

   ____

7. Set an ohmmeter on the R × 1 scale and adjust the meter for a value of 0 Ω when the leads are shorted together. Place the ohmmeter leads across terminals 1 and 2 on the direct current machine module. Slowly rotate the armature by hand and observe the action of the ohmmeter. Does the ohmmeter reading remain constant, or does it become erratic?

   ____

8. Is it possible to measure an accurate value of resistance?

   ____

9. Reconnect the ohmmeter leads across terminals 3 and 4 on the module. Slowly rotate the armature by hand. Does the ohmmeter reading remain constant, or does it become erratic?

   ____

10. Measure the resistance of the winding.

    ____________ Ω

11. Reconnect the ohmmeter leads across terminals 5 and 6 of the module. Slowly turn the armature by hand. Does the ohmmeter reading remain constant, or does it become erratic?

    ____

12. Measure the resistance of the field.

    ____________ Ω

This ohmmeter test can be used to determine which leads are connected to the armature, series field, and shunt field of a DC machine. The armature is connected across terminals 1 and 2. The erratic reading of the ohmmeter is caused by the brushes making and breaking contact with different sets of coils wound around the armature. It would be very difficult to determine the resistance of the armature using an ohmmeter. To determine the armature resistance, a different method must be used.

The series field is connected to terminals 3 and 4. This is determined by the fact that the ohmmeter reading remains constant when the armature is rotated by hand and by the fact that it has a very low resistance.

The shunt field is connected to terminals 5 and 6. The ohmmeter reading remained constant when the armature was turned by hand, and it has a relatively high resistance.

13. Reconnect the ohmmeter leads to terminals 7 and 8. Rotate the shunt field rheostat control knob fully counterclockwise. Measure the resistance across terminals 7 and 8.

    ______________ Ω

14. Rotate the control knob fully clockwise. Measure the resistance between terminals 7 and 8.

    ______________ Ω

15. Insert the direct current machine module into the mobile console beside the EMS power supply.
16. Insert the electrodynamometer module (EMS 8911), Figure 24-3, or the primemover/dynamometer module (EMS 8960) beside the direct current machine. If you are using the primemover/dynamometer module, it will be necessary to make the 24 VAC power connection to the power supply.
17. Connect the timing pulley of the direct current machine to the timing pulley of the electrodynamometer with the timing belt. Be sure to place the timing belt between the two spring-loaded bearings. The bearings are used to provide constant belt tension.
18. Connect the circuit shown in Figure 24-4.
19. Check to see if the voltage control knob of the power supply is in the full counterclockwise (0) position.
20. Turn on the power supply and adjust the voltage until there is a current of 2.5 A flowing in the circuit.
21. Measure the voltage drop across the series field.

    ______________ V

22. Compute the resistance of the series field using the formula

$$R = \frac{E}{I}$$

    ______________ Ω

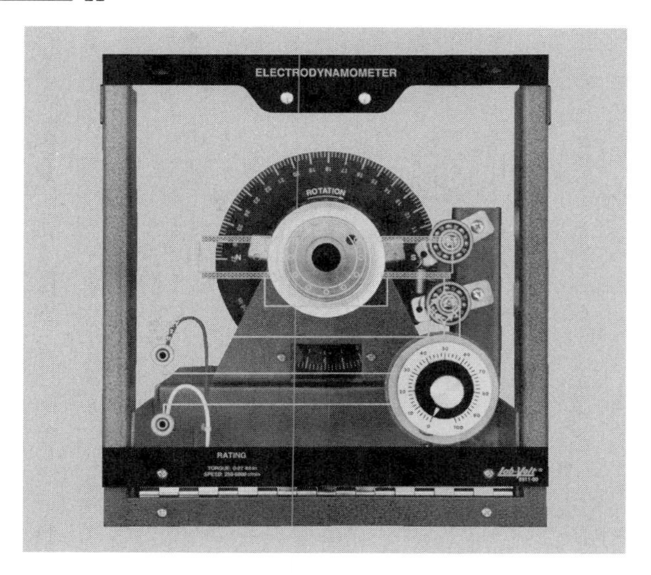

FIGURE 24-3 Electrodynamometer module (Courtesy of Lab-Volt® Systems, Inc.)

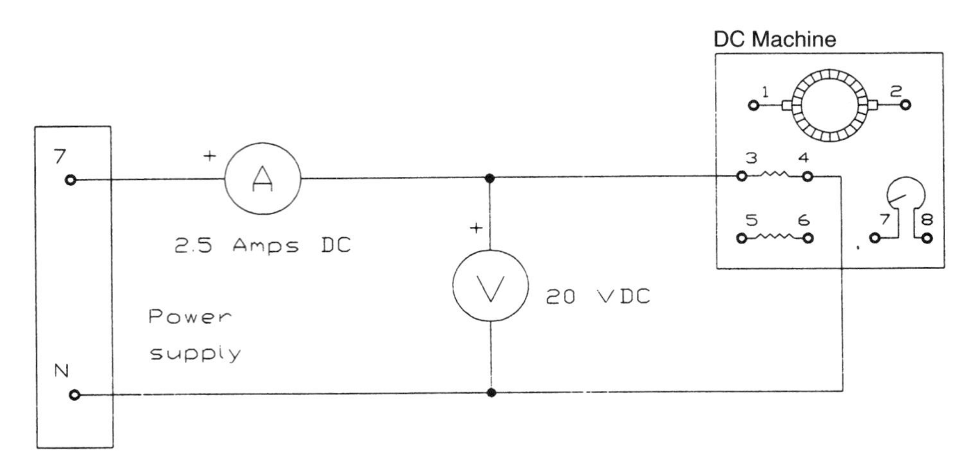

FIGURE 24-4 Testing the series field

23. Compare this resistance value with the value found in step 10.
24. **Return the voltage to 0 V and turn off the power supply.**
25. Connect the circuit shown in Figure 24-5.
26. Turn on the power supply and adjust the voltage until a current of 0.4 A flows in the circuit.
27. Measure the voltage drop across the shunt field.

    _____________ V

28. Compute the resistance of the shunt field using the formula

$$R = \frac{E}{I}$$

    _____________ Ω

29. Compare this value with the value found in step 12.
30. **Return the voltage to 0 V and turn off the power supply.**

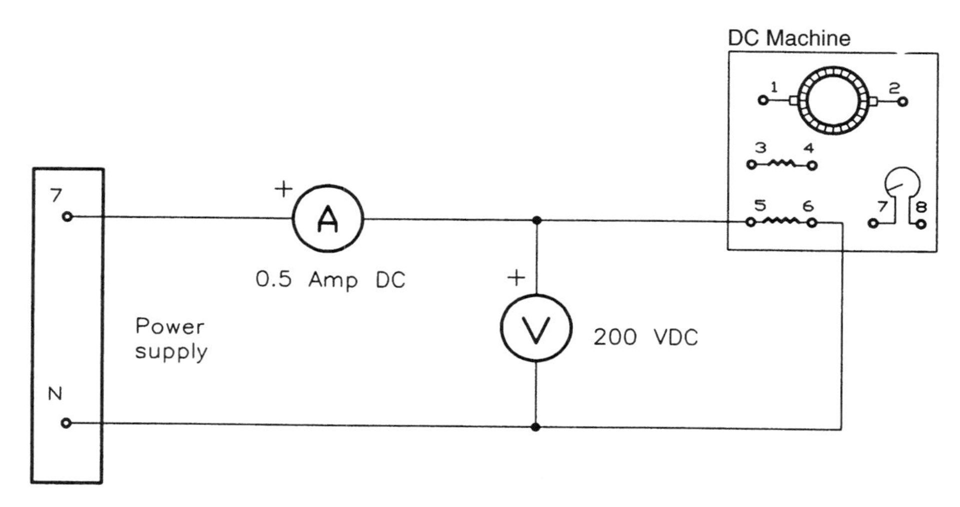

FIGURE 24-5 Testing the shunt field

31. Connect the circuit shown in Figure 24-6. The circuit illustrates the connection of the EMS 8911 electrodynamometer. If you are using the EMS 8960 prime mover/dynamometer, it is necessary to make only the 24 VAC connection to the power supply.
32. Adjust the shunt field rheostat control, located on the direct current machine module, to the full counterclockwise position.
33. Turn on the power supply.
34. Adjust the electrodynamometer control knob to the full clockwise position.
35. Slowly adjust the voltage control knob on the power supply until the direct current machine begins to turn.
36. Measure the amount of current flow through the armature and the voltage drop across the armature.

    ______________ A ______________ V

37. **Return the voltage control to 0 V and turn off the power supply.**
38. Return the electrodynamometer control to the full counterclockwise position.
39. Compute the resistance of the armature using the voltage and current values found in step 36.

    ______________ Ω

    The resistance of the armature is measured in this way because it gives a true average value. When the armature is turning at an extremely slow rate, almost no counter-voltage is induced into the armature and the current flow is limited by the resistance of the armature.

40. Disconnect the circuit and return the equipment to its proper place.

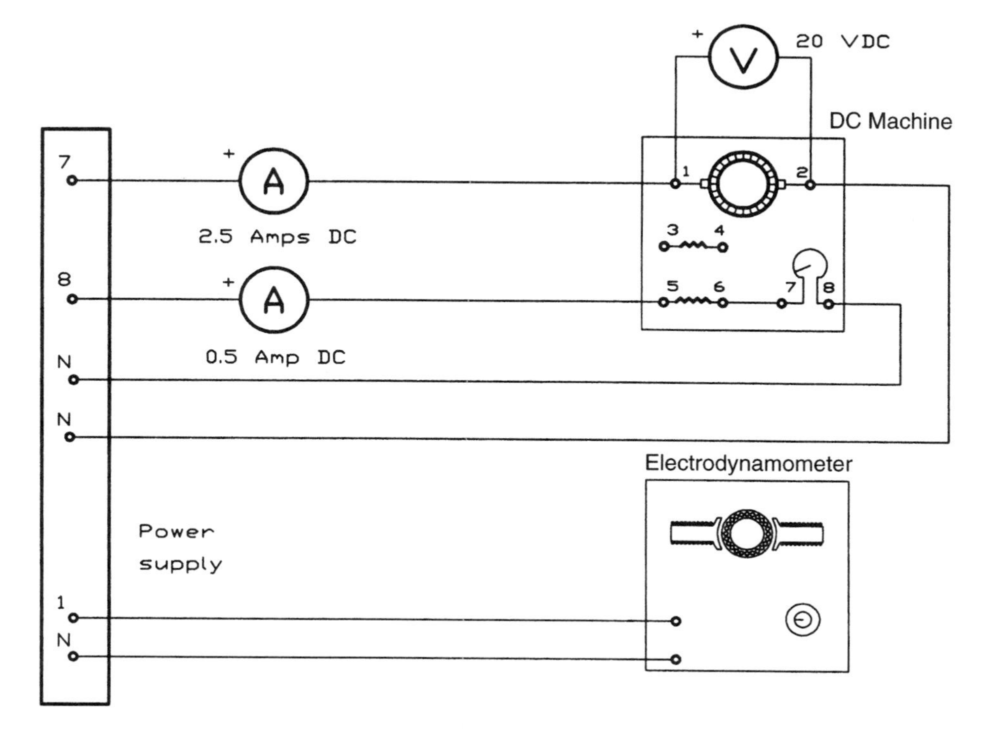

FIGURE 24-6 Determining the armature resistance

Review Questions

1. What is the rotating member of a direct current machine called?

   ____

2. What are the characteristics of a series field?

   ____

3. What are the characteristics of a shunt field?

   ____

4. Is the resistance of the shunt field higher or lower than the resistance of the series field?

   ____

5. What are the functions of the commutator?

   ____

6. Explain how to locate the armature leads with an ohmmeter.

   ____

7. Why are brushes made of a soft carbon material?

   ____

Exercise 25

Setting the Neutral Plane of a Direct Current Machine and Residual Magnetism of the Pole Pieces

Objectives

After completing this lab you should be able to:

- Define the neutral plane of a direct current machine.
- Set the neutral plane of a direct current machine.

Materials and Equipment

Power supply module	EMS 8821
Direct current machine module	EMS 8211
AC voltmeter module	EMS 8426
AC ammeter module	EMS 8425

Discussion

The neutral plane of a direct current machine is the point on the commutator where the armature windings are parallel to the magnetic flux lines of the pole pieces. When the armature windings are parallel to the flux lines, there is no cutting action and, therefore, no induced voltage. The neutral plane is the position at which the brushes should be set. If the brushes are not set at the neutral plane, the brushes will make and break contact with the commutator bars when there is a potential or voltage across the bars. This will cause arcing and sparking at the brushes. The arcs and sparks produce heat, which can damage the brushes and commutator. Poor commutation also causes a power loss in the machine and reduced efficiency.

The simplest method of setting the brushes to the neutral plane position is to apply alternating current to the armature and connect an AC voltmeter across the shunt field windings. When alternating current flows through the armature, the windings of the armature act as the primary of a transformer. The shunt field windings act as the secondary of a transformer. If the brushes are not set in the neutral plane position, the magnetic field of the armature will induce a voltage into the shunt field winding. The AC voltmeter connected across the shunt field winding will measure the amount of voltage induced.

When using this method of setting the neutral plane, be careful not to permit too much current to flow though the armature. A variable voltage AC source or a current-limiting resistor and AC ammeter are needed.

RESIDUAL MAGNETISM

Series and shunt field windings of a direct current machine wound around metal pole pieces and are used to produce a magnetic field in the pole piece, as in Figure 25-1.

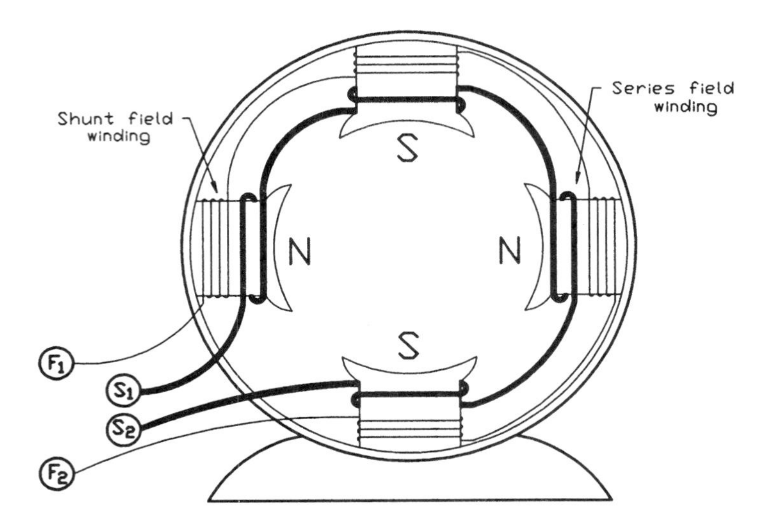

FIGURE 25-1 Both series and shunt field windings are contained on each pole piece.

Because the power applied is DC, the magnetic field produced never changes polarity. The molecules of metal in the pole pieces align themselves in the direction of the magnetic field.

When the DC power is disconnected from the pole pieces, most of the metal molecules return to a neutral position (demagnetize), but some remain in a magnetized position. This produces some amount of residual magnetism in the pole pieces. This residual magnetism is used to produce an initial output voltage in some DC generators. Generators that use residual magnetism to produce an initial voltage are called self-excited. These machines will not operate without some amount of residual magnetism left in the pole pieces.

Other types of generators, called separately excited, use an external power source to produce the magnetic field in the pole pieces. Residual magnetism to these machines is neither desired nor wanted.

Regardless of the type of machine being used, the amount of residual magnetism left in the pole pieces is often determined by the conditions of the machine when it was last used. This means that the amount of residual magnetism left in the pole pieces can be different at the beginning of each experiment and can, therefore, cause a wide range in answers from one experiment to the next. The following procedures can be used either to remove the residual magnetism before beginning an experiment or to set the residual magnetism to a certain level.

Removing Residual Magnetism

The residual magnetism can be removed by applying strong AC current to the pole pieces and then slowly removing it. Because the current is alternating, it will cause the magnetic field to reverse directions 60 times per second. If the current is slowly decreased, the molecules in the pole pieces will be left in a state of disarray, effectively demagnetizing the pole piece. To remove the residual magnetism follow the procedure described below:

1. Connect the circuit shown in Figure 25-2.
2. Set the voltage control knob to the full counterclockwise position (0 V).
3. Turn on the power supply and increase the voltage until a current of 2.5 A flows through the series field.
4. Slowly return the voltage to 0 V.

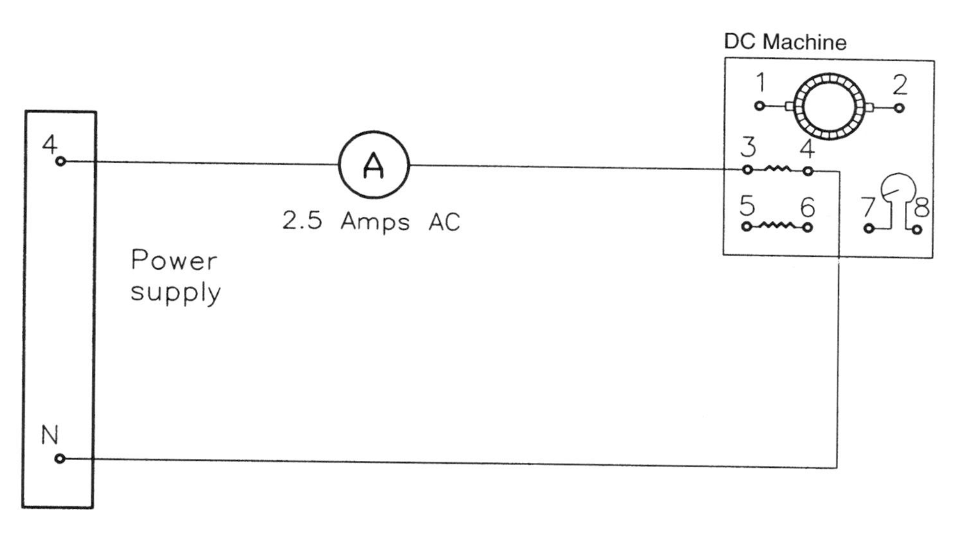

FIGURE 25-2 Removing the residual magnetism

5. **Turn off the power supply and disconnect the circuit. The pole pieces have now been demagnetized.**

Setting the Amount of Residual Magnetism

It may become desirable to preset the pole pieces so they will have a certain amount of residual magnetism before beginning an experiment. This would produce more uniform results when working with self-excited generators. The following procedure can be used to preset the amount of residual magnetism in the pole pieces.

1. Connect the circuit shown in Figure 25-3.
2. Set the voltage control knob to the full counterclockwise position (0 V).
3. Turn on the power supply and increase the voltage until a current of 0.4 A DC flows through the shunt field winding.
4. **Return the voltage to 0 V and turn off the power supply.**
5. Disconnect the circuit. The residual magnetism has now been set.

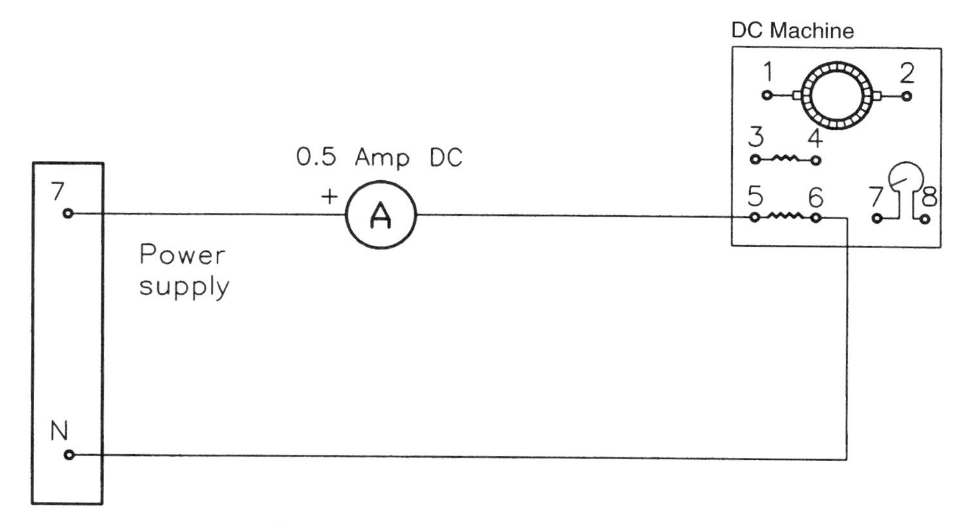

FIGURE 25-3 Setting the residual magnetism

Name ______________________________ Date ______________

Procedure

1. Connect the circuit shown in Figure 25-4.
2. Turn on the power supply and adjust the voltage until a current of 0.5 A AC flows through the armature.
3. Open the front cover of the direct current machine module.
4. Move the brush control position lever back and forth several times and observe the AC voltmeter connected to the shunt field.
5. Does the voltage increase and decrease as the lever is moved?

   ____

6. Set the brush control lever to the position where the amount of voltage induced into the shunt field is at the lowest value. This should be 0 V or very close to it. The brushes have now been set in the neutral plane position.
7. Close the front cover of the direct current machine module.
8. **Return the voltage setting on the power supply to 0 V and turn off the power supply.**
9. Disconnect the circuit and return the components to their proper place.

Review Questions

1. What is the neutral plane position of a DC machine?

   ____

2. What are some effects of the brushes not being set in the neutral plane position?

   ____

3. Explain why connecting AC voltage to the armature of a DC machine causes voltage to be induced into the shunt field.

   ____

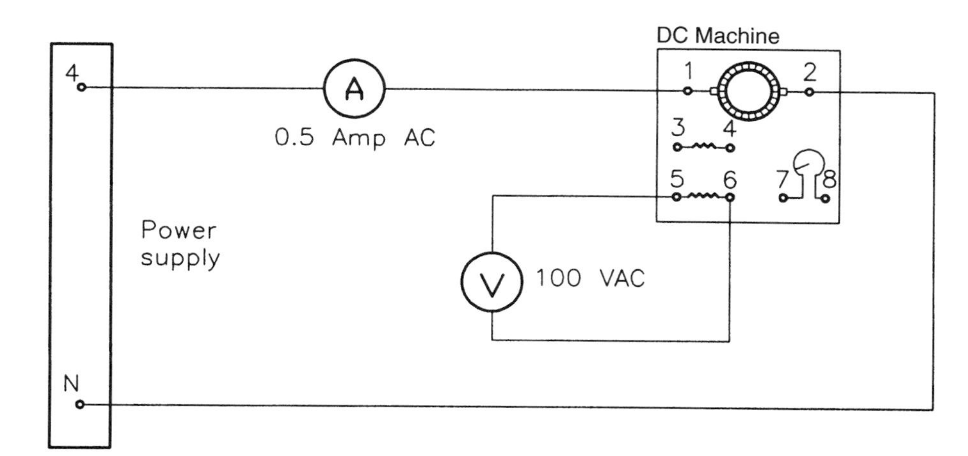

FIGURE 25-4 Setting the neutral plane of a direct current machine

Exercise 26

Series Generators

Objectives

After completing this lab you should be able to:

- Discuss the operation of a series connected DC generator.
- Connect a DC generator for operation as a series machine and make electrical measurements.

Materials and Equipment

Power supply module	EMS 8821
DC metering module	EMS 8412
AC ammeter module	EMS 8425
Direct current machine module	EMS 8211
Synchronous machine module	EMS 8241
or prime mover/dynamometer module	EMS 8960
Variable-resistance module	EMS 8311

Discussion

The series generator is so named because it contains only a series field connected in series with the armature. Any current produced by the generator must flow through the series field. This causes an increase in magnetic field strength in the pole pieces and a corresponding increase in output voltage. Three factors determine the output voltage of a DC generator:

1. the number of turns of wire in the armature,
2. the speed of the cutting action (speed of rotation), and
3. the strength of the magnetic field in the pole pieces.

If any of these factors change, the output voltage of the generator will change also.

In this experiment, it will be seen that the output voltage will increase each time a load resistor is added. Bear in mind that each time a load resistor is added in parallel, total load resistance decreases. This decrease of resistance causes an increase in output current each time a resistor is added. The increase of load current produces a stronger magnetic field in the pole pieces, which causes an increase in the output voltage.

Although the voltage will increase, the amount of increase will be small. The reason for this is that the minimum resistance value that can be obtained by the variable-resistance module is 57.1 Ω. This value of load resistance is too great to permit the amount of current flow needed to produce a dramatic increase in output voltage. It does, however, illustrate the basic operating characteristic of a series generator, which is to increase output voltage with an increase of load current. It is possible to add external resistance to the load circuit to reduce the total load resistance, increasing the output current and voltage. This should not be done without the supervision of an instructor, however, because it is possible to damage the DC machine because of excessive current flow.

Name ______________________________ Date ______________

Procedure

1. Open the face plate of the direct current machine and the synchronous machine (shown in Figure 26-1) or the prime mover/dynamometer, depending on which machine is being used. The prime mover/dynamometer may be employed in place of the synchronous machine.

2. Connect the two machines together with the timing belt. Be sure the timing belt is between the spring-loaded bearings.

3. Connect the circuit shown in Figure 26-2 if the synchronous machine is being used as the prime mover in this experiment. If the prime mover/dynamometer is being used, connect the circuit shown in Figure 26-3.

4. If the synchronous machine module is being used, open (turn off) switch S1 on the synchronous machine's module. Open all the switches on the variable resistance module. The synchronous machine will provide a constant speed of 1800 RPM (revolutions per minute) and is used to supply the turning force for the DC machine. Switch S1 is used to supply direct current to the rotor of the synchronous machine. This is referred to as **exciting** the rotor.

   CAUTION: *The synchronous machine should never be started with switch S1 in the closed or on position. Switch S1 should be closed only after the motor is running.*

   If the EMS 8960 prime mover/dynamometer is being used instead of the synchronous machine, it should be set for a speed of 1800 RPM and maintained throughout the experiment. Make certain the MODE switch is set in the prime mover position and the DISPLAY switch is set in the speed position.

5. Momentarily turn on the power and observe the direction of rotation of the synchronous machine and DC generator. The machine should turn in the clockwise direction as you face the machine. If it does not, switch the wires connected to terminals 1 and 2 of the synchronous machine. This will cause the direction of rotation to reverse.

FIGURE 26-1 Synchronous machine module.
(Courtesy of Lab-Volt® Systems, Inc.)

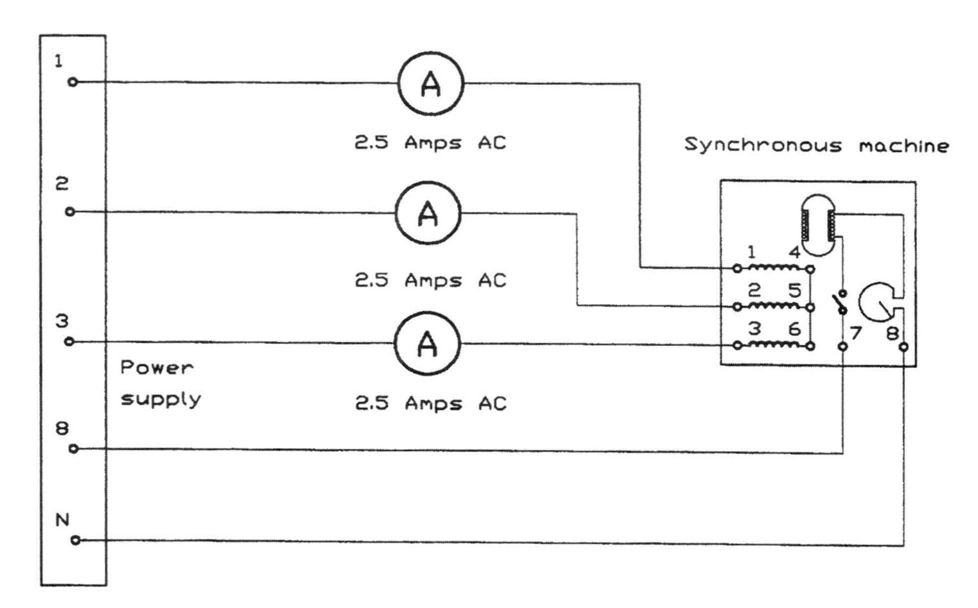

FIGURE 26-2 Connecting the AC synchronous machine

NOTE: It may be desirable to preset the amount of residual magnetism in the pole pieces of the DC machine before connecting the series generator. To do this, follow the procedure for setting the amount of residual magnetism described in Exercise 25.

6. Connect the circuit shown in Figure 26-4. Do not disconnect the synchronous machine or the prime mover/dynamometer.
7. Turn on the power supply and close switch S_1 on the synchronous machine module, or adjust the prime mover for 1800 RPM.
8. If the synchronous machine is being used, adjust the control knob on the front of the synchronous machine module until the three AC ammeters indicate the lowest value of current.
9. Measure the amount of current flow from the DC generator to the variable-resistance module.

   ______________ A

10. Measure the amount of DC voltage being applied to the variable-resistance module.

    ______________ V

    The voltage being applied to the variable-resistance module is produced because of residual magnetism in the pole pieces. Generators that use residual magnetism to produce an initial output voltage are known as self-excited generators. Generators that have no residual magnetism in the pole pieces and must depend on an outside current source to produce a magnetic field in the pole pieces are known as separately excited generators.

11. Close the switches necessary on the variable-resistance module to produce 171.4 Ω.
12. Measure the amount of current flow from the DC generator to the load.

    ______________ A

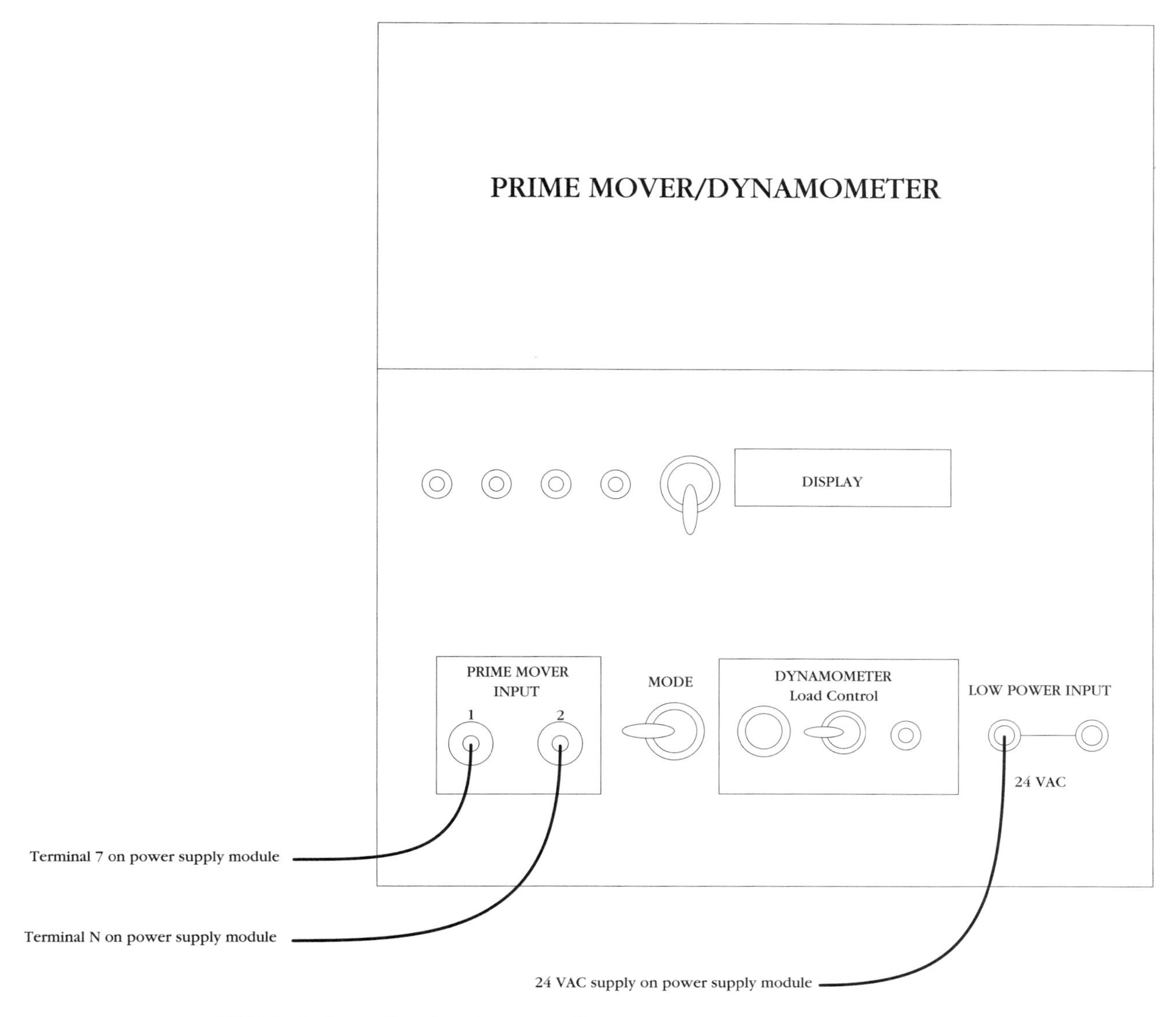

FIGURE 26-3 Connecting the prime mover/dynamometer module.

13. Measure the amount of DC voltage being applied to the variable-resistance module.

    ______________ V

14. Close the switches necessary on the variable-resistance module to produce a resistance of 85.7 Ω.

15. Measure the amount of current flow from the DC generator to the load.

    ______________ A

16. Measure the amount of DC voltage being applied to the variable-resistance module.

    ______________ V

17. Close the switches necessary on the variable-resistance module to produce a resistance of 57.1 Ω.

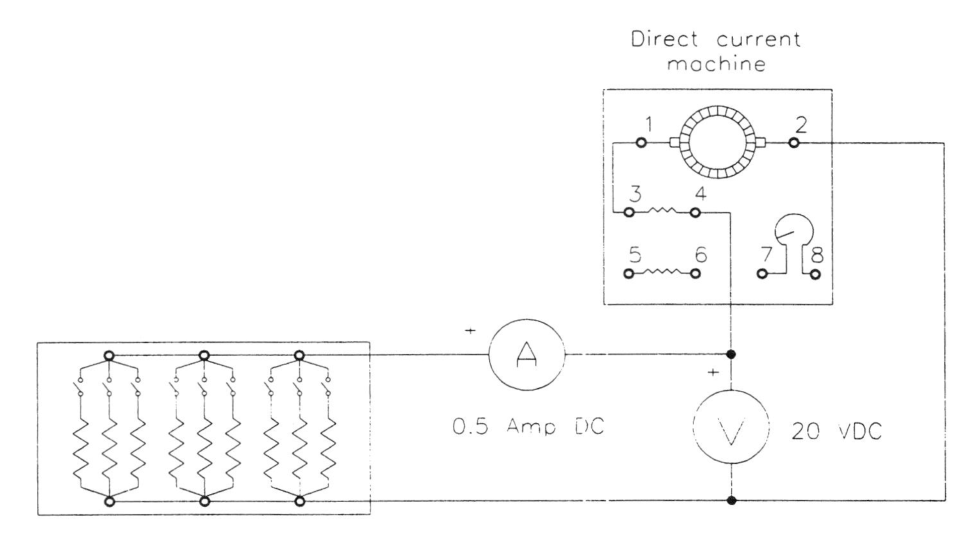

FIGURE 26-4 Series generator connection

18. Measure the amount of current flow from the DC generator to the load.

    ____________ A

19. Measure the amount of DC voltage being applied to the variable-resistance module.

    ____________ V

20. **Open all switches on the variable resistance module and turn off the power supply.**

21. Disconnect the circuit and return the components to their proper place.

Review Questions

1. What three factors determine the output voltage of a DC generator?

   A. ____

   B. ____

   C. ____

2. Why does the output voltage of a series-connected DC generator increase with an increase of load current?

   ____

3. What caused the initial voltage to be produced when the generator was first started before load was added to the circuit?

   ____

4. What would have happened when load was added to the circuit if this initial output voltage had not been present?

   ____

5. What is a self-excited generator?

   ____

Exercise 27

Separately Excited Direct Current Shunt Generators

Objectives

After completing this lab you should be able to:

- Discuss the operation of a shunt generator.
- Connect a shunt generator and make measurements using measuring instruments.
- Draw a characteristic curve of a shunt generator.

Materials and Equipment

Power supply module	EMS 8821
DC metering module	EMS 8412
AC ammeter module	EMS 8425
Direct current machine module	EMS 8211
Synchronous machine module	EMS 8241
Variable-resistance module	EMS 8311

Discussion

The shunt generator has the shunt field connected in parallel with the armature. This permits the amount of current flow through the field to be controlled separately from the current flow through the armature, as in Figure 27-1. A separately excited generator, however, has the shunt field connected to a separate power supply, as shown in Figure 27-2. This permits a greater degree of control over the output voltage. Recall that the output voltage of a generator is determined by three factors:

1. the number of turns of wire in the armature,
2. the speed of rotation of the armature, and
3. the strength of the magnetic field in the pole pieces.

Because the shunt file in this machine produces the magnetic field in the pole pieces, the amount of current flow through the field will determine the strength of the field.

A true separately excited generator will not have residual magnetism in its pole pieces. This lack of residual magnetism permits the output voltage to be adjusted to 0 V if desired. The direct current machine used in this experiment, however, does contain residual magnetism in its pole pieces, and for this reason it will not be possible to adjust the output voltage to 0 V.

The three-phase AC synchronous machine will be used to provide the turning force for the DC generator because it will not decrease its speed under load. When a generator produces power, it develops countertorque. This means that the generator is harder to turn. The amount of countertorque produced is proportional to the amount of power (watts) the generator is producing. Countertorque is caused by the magnetic field of the armature being attracted to the magnetic field of the pole pieces. The more current produced in the armature, the greater the countertorque becomes.

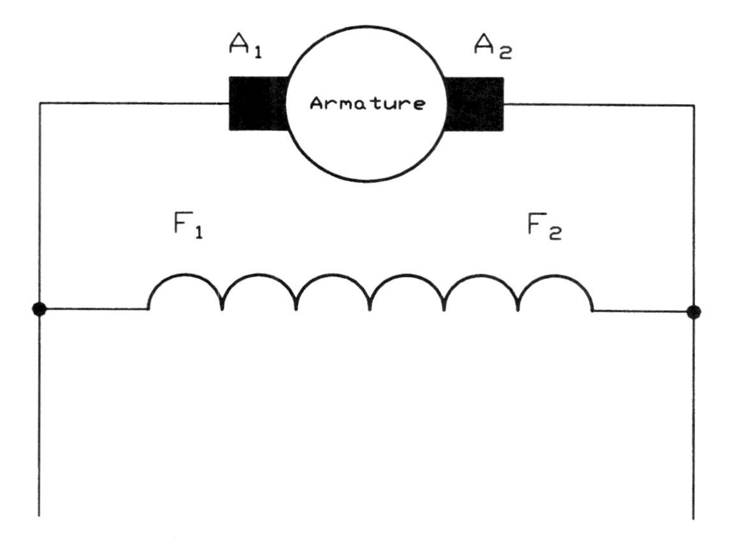

FIGURE 27-1 Shunt generator

During the experiment, the effect of countertorque can be seen on the AC ammeters connected in series with the synchronous machine. The current flow to the synchronous machine will increase when the load on the DC generator increases.

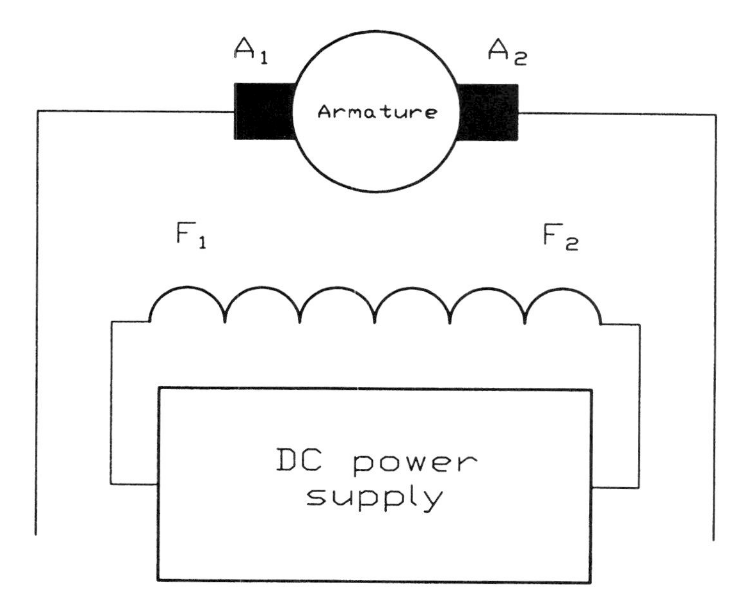

FIGURE 27-2 Shunt field connected to an external power source

In this experiment, the output voltage will decrease as generator load is increased. Several factors cause this decrease of voltage. Four of them are discussed below:

ARMATURE REACTION

Armature reaction is the twisting or bending of the main magnetic field of the pole pieces. It is caused by the magnetic field produced in the armature when load is added to the machine. The turning armature magnetic field distorts the stationary magnetic field of the pole pieces. The amount of armature reaction is proportional to the load current of the generator.

EDDY CURRENT LOSS

Eddy currents are induced into the iron core material of the generator itself. As the armature spins through the magnetic field of the pole pieces, current is induced into the core material. Core material is generally laminated to help reduce eddy current induction.

HYSTERESIS LOSS

Hysteresis loss is caused by the reversal of current flow in the armature. Each time the direction of current flow in the armature changes, the molecules of iron turn around to realign themselves with the new magnetic field polarity.

ARMATURE RESISTANCE

The greatest cause of voltage drop is the resistance of the wire in the armature winding. The lower the armature resistance, the less voltage drop the generator will have. For example, assume that the armature has a resistance of 10 Ω. Now assume that there is a current flow of 3 A. The amount of voltage drop can be found by using Ohm's law.

$$E = I \times R$$
$$E = 3\ A \times 10\ \Omega$$
$$E = 30\ V$$

Now assume that the armature has a resistance of 2 Ω. If 3 A of current flow is produced by this armature, the voltage drop will be

$$E = I \times R$$
$$E = 3\ A \times 2\ \Omega$$
$$E = 6\ V$$

The lower the armature resistance, the less the output voltage drop will be. The voltage regulation of a DC generator is proportional to the armature resistance.

Name ______________________________ Date ______________

Procedure

1. Open the face plate of the direct current machine and the synchronous machine.
2. Connect the two machines together with the timing belt. Be sure that the timing belt is between the spring-loaded bearings.
3. Connect the circuit shown in Figure 27-3.
4. Open (turn off) switch S_1 on the synchronous machine module and all the switches on the resistance load module. Switch S_1 on the synchronous machine module is used to apply direct current to the rotor of the AC motor. This is referred to as exciting the rotor.

   Caution: ***The synchronous machine should never be started with switch S_1 in the closed position.***

   Switch S_1 should be close only after the synchronous motor is running.
5. Momentarily turn on the power and observe the direction of rotation of the synchronous machine and DC generator. The machines should turn in the clockwise direction as you face them. If they do not, turn off the power and switch the wires connected to terminals 1 and 2 of the synchronous machine. This will cause the direction of rotation to reverse.

   NOTE: It may be desirable to demagnetize the pole pieces of the DC machine before making connection. Follow the procedure described in Exercise 25 for removing residual magnetism.
6. Adjust the voltage control knob on the power supply to the full counterclockwise position.

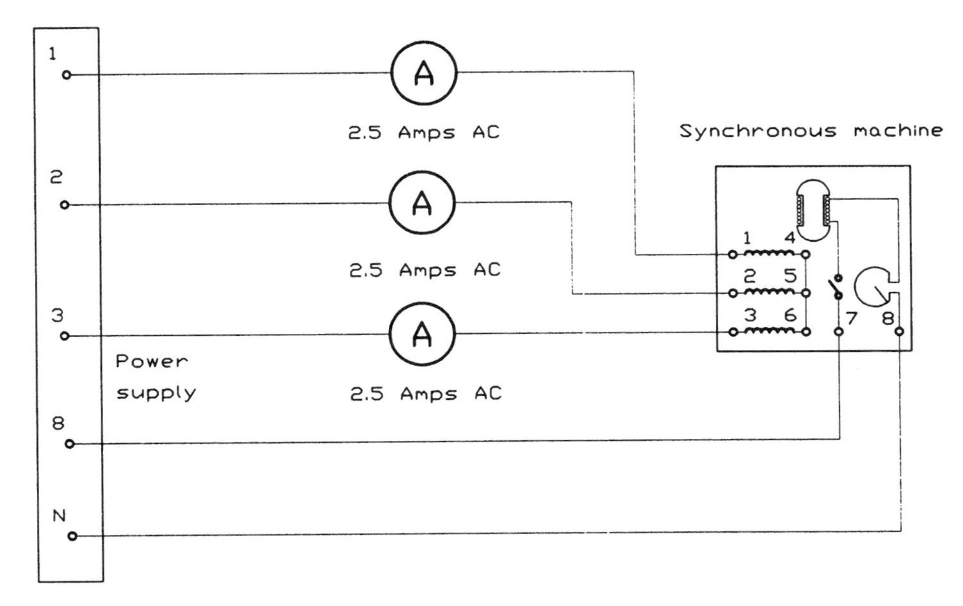

FIGURE 27-3 Synchronous motor connection

7. Connect the circuit shown in Figure 27-4. Do not disconnect the synchronous machine.
8. Turn on the power supply and close switch S_1 on the synchronous machine module.
9. Adjust the control knob on the front of the synchronous machine module until the three AC ammeters indicate the lowest value of current.
10. Record the value of output voltage and the amount of shunt field current for the DC machine.

    Output voltage = ______________ V   Field current = ______________ A

11. Increase the amount of shunt field current by increasing the DC voltage applied to it. This can be adjusted with the control knob located on the power supply. Increase the shunt field current until the output voltage is 80 V. Measure the amount of shunt field current at this point.

    Field current = ______________ A

12. Increase the shunt field current until the output voltage has a value of 100 V. Measure the shunt field current at this point.

    Field current = ______________ A

13. Increase the shunt field current until the output voltage has a value of 120 V. Measure the shunt field current at this point.

    Field current = ______________ A

14. Fill in the chart in Figure 27-5 by changing the resistance values of the variable-resistance module. Do not readjust the shunt field control for the remainder of this experiment.
15. **Turn off the power supply and disconnect the circuit. Return the components to their proper place.**
16. Using the measured values from the chart in Figure 27-5, draw a characteristic curve for this generator. This is done by plotting the points for output voltage and output current on the graph shown in Figure 27-6. When the points have been plotted, draw through the points with a continuous line.

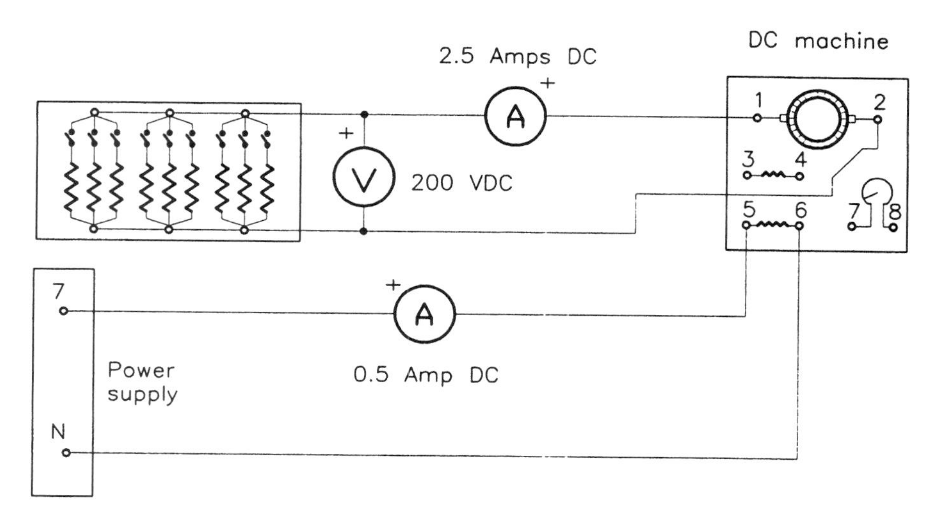

FIGURE 27-4 Separately excited shunt generator connection

Resistance (ohms)	Output Voltage (volts)	Output Current (amps)	AC current (amps)
Infinity			
1200			
600			
400			
300			
240			
150			
100			
80			
57.1			

FIGURE 27-5

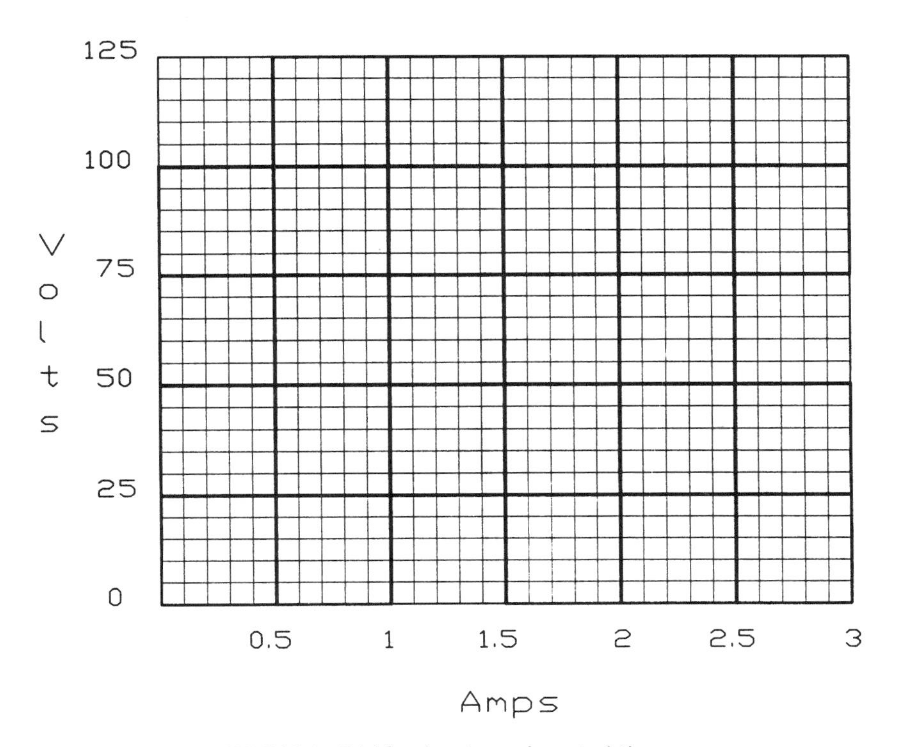

FIGURE 27-6 Grid for drawing a characteristic curve

Review Questions

1. What three factors determine the output voltage of a generator?

   A. ______

   B. ______

   C. ______

2. Is the shunt field connected to the armature in this experiment?

   ______

3. What is a separately excited generator?

   ______

4. Explain why an increase of shunt field current causes an increase in the output voltage.

   ______

5. Why did the AC amperage to the synchronous machine increase as load was added to the generator?

   ______

6. Define the following losses:

   A. Armature reaction ______

   B. Eddy current ______

   C. Hysteresis loss ______

7. An armature has a resistance of 5 Ω. What will be the voltage drop across the armature when 5 A of current flows through it?

   ______ V

8. What is the voltage regulation of a DC generator proportional to?

   ______

Exercise 28

Self-Excited Direct Current Shunt Generators

Objectives

After completing this lab you should be able to:

- Discuss the operation of a self-excited shunt generator.
- Connect a self-excited shunt generator and make measurements using measuring instruments.
- Draw a characteristic curve of a self-excited shunt generator.

Materials and Equipment

Power supply module	EMS 8821
DC metering module	EMS 8412
AC ammeter module	EMS 8425
Direct current machine module	EMS 8211
Synchronous machine module	EMS 8241
Variable-resistance module	EMS 8311

Discussion

The self-excited shunt generator has the shunt field connected in parallel with the armature. In the self-excited generator the voltage produced by the armature is used to furnish current for the shunt field. The current flow through the shunt field is, therefore, dependent on the amount of voltage produced by the armature.

When the generator is first started, residual magnetism in the pole pieces is used to produce a small amount of voltage. Because the shunt field is connected in parallel with the armature, as in Figure 28-1, current begins to flow through the shunt field. This causes an increase in the magnetic field strength of the pole pieces, which causes a higher voltage to be produced in the armature. Armature voltage and shunt field current will continue to increase until a maximum value is reached. This maximum value is determined by several factors, such as the resistance of the armature, the resistance of the shunt field winding, and the speed of rotation of the armature.

The output voltage of the generator is controlled by the strength of the magnetic field of the pole pieces. Several methods can be used to obtain this control. In large commercial installations, the shunt field current is controlled by electronic regulators. The old-style automotive generators were controlled by electromechanical regulators. The Lab-Volt EMS system uses a shunt field rheostat to control the shunt field current. The shunt field rheostat is connected in series with the shunt field. This permits resistance to be added to or removed from the shunt field circuit and thereby control the current flow.

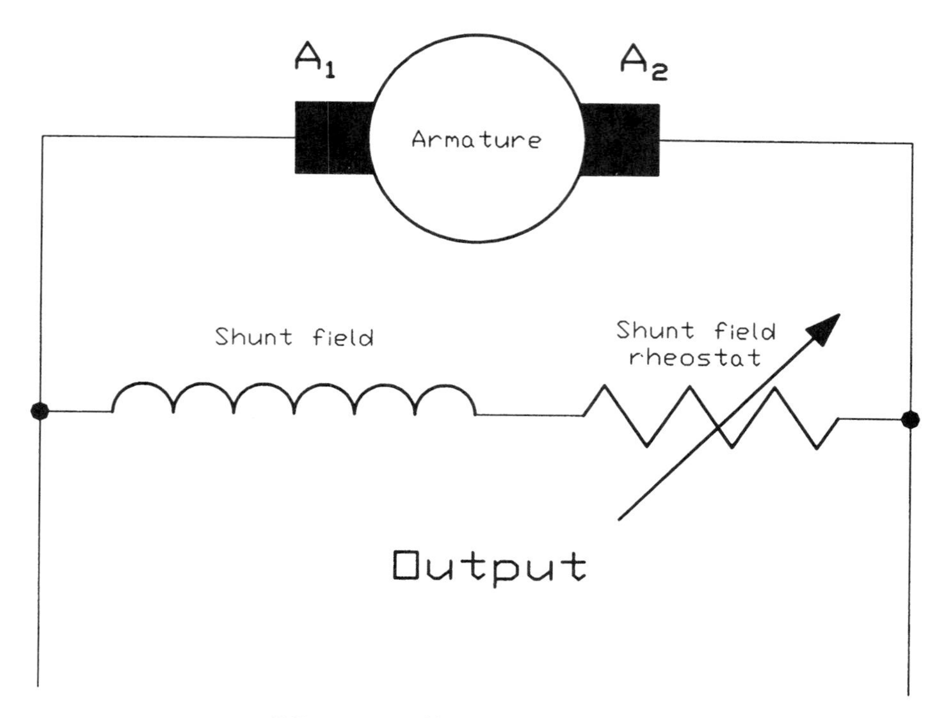

FIGURE 28-1 Self-excited shunt generator

The self-excited generator will exhibit a greater voltage drop than the separately excited generator. The reason for this is that the self-excited generator depends on the armature voltage to provide the current flow through the field. When the armature voltage drops, shunt field current will decrease and weaken the magnetic field strength. The weaker magnetic field will cause a greater reduction of armature voltage. The separately excited generator does not have this problem because its shunt field current is furnished by a source that is not dependent on armature voltage.

The output voltage of a self-excited generator can be controlled by reducing the resistance of the shunt field rheostat and permitting more shunt field current to flow. The purpose of this experiment, however, is to illustrate the characteristic curve of a self-excited shunt generator. Once the no-load voltage has been adjusted to the desired value by the shunt field rheostat, it will not be adjusted again for the remainder of the experiment.

Name ______________________________ Date ______________

Procedure

1. Open the face plate of the direct current machine and the synchronous machine or the prime mover/dynamometer, depending on which machine is being used. The prime mover/dynamometer may be employed in place of the synchronous machine.
2. Connect the two machines together with the timing belt. Be sure the timing belt is between the spring-loaded bearings.
3. Connect the circuit shown in Figure 28-2 if the synchronous machine is being used as the prime mover in this experiment. If the prime mover/dynamometer is being used, connect the circuit shown in Figure 28-3.
4. If the synchronous machine module is being used, open (turn off) switch S1 on the synchronous machine's module. Open all the switches on the variable resistance module. The synchronous machine will provide a constant speed of 1800 RPM (revolutions per minute) and is used to supply the turning force for the DC machine. Switch S1 is used to supply direct current to the rotor of the synchronous machine. This is referred to as **exciting** the rotor.

   CAUTION: *The synchronous machine should never be started with switch S1 in the closed or on position. Switch S1 should be closed only after the motor is running.*

   If the EMS 8960 prime mover/dynamometer is being used instead of the synchronous machine, it should be set for a speed of 1800 RPM and maintained throughout the experiment. Make certain the MODE switch is set in the prime mover position and the DISPLAY switch is set in the speed position.
5. Momentarily turn on the power and observe the direction of rotation of the synchronous machine and DC generator. The machines should turn in the clockwise direction as you face them. If they do not, switch the wires connected to terminals 1 and 2 of the synchronous machine.

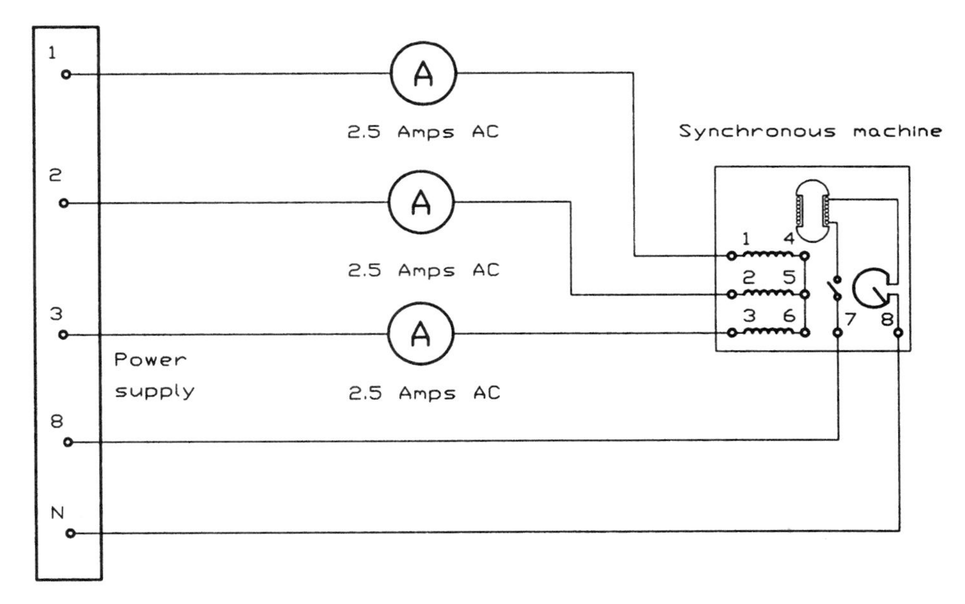

FIGURE 28-2 Synchronous motor connection

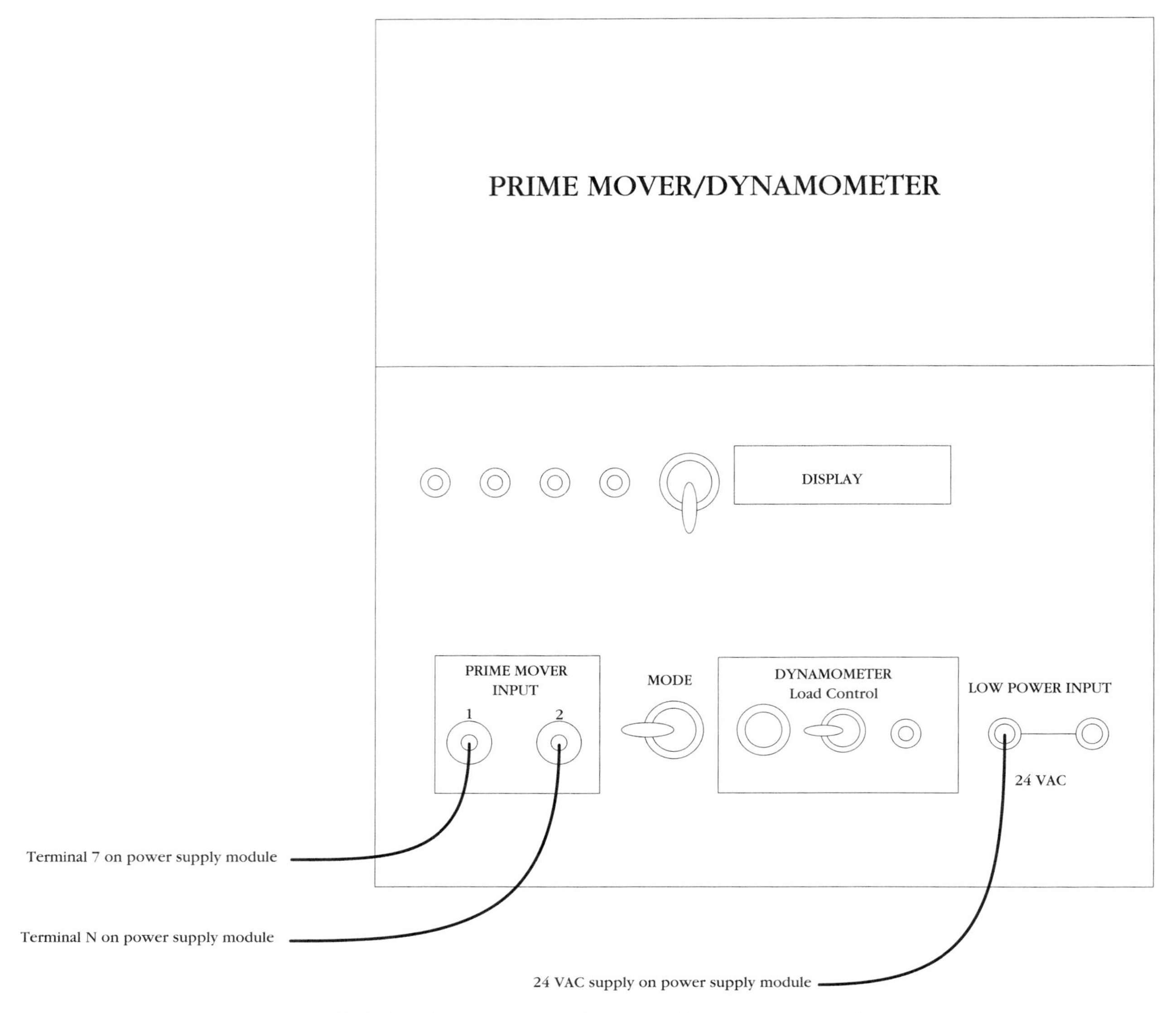

FIGURE 28-3 Connecting the prime mover/dynamometer module.

NOTE: It may be desirable to set the residual of the pole pieces before connecting the DC machine. Follow the procedure described in Exercise 25 for setting the amount of residual magnetism.

6. Adjust the voltage control knob on the power supply to the full counterclockwise position.
7. Connect the circuit shown in Figure 28-4.
8. Adjust the shunt field rheostat control knob located on the direct current machine module to the full counterclockwise position.
9. Turn on the power supply and close switch S_1 on the synchronous machine module, or set the prime mover for 1800 RPM.
10. Adjust the control knob on the front of the synchronous machine module until the three AC ammeters indicate the lowest value of current.

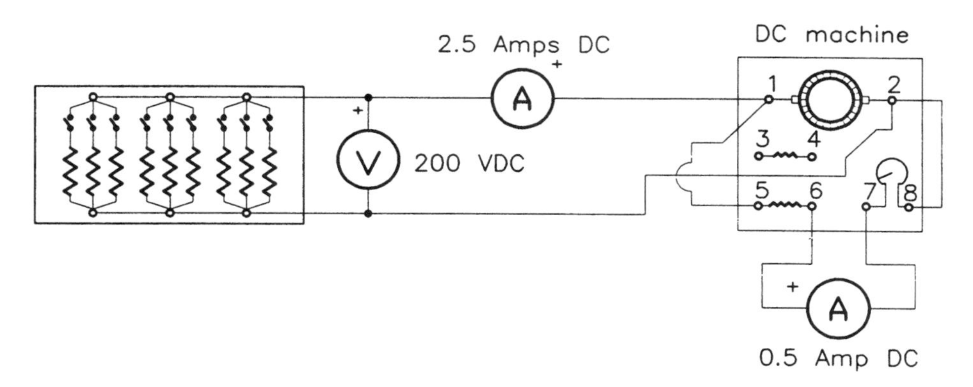

FIGURE 28-4 Self-excited shunt generator connection

11. Adjust the shunt field rheostat located on the direct current machine module until the generator produces a voltage of 120 V.
12. Fill in the chart in Figure 28-5 by changing the resistance values of the variable-resistance module. Do not readjust the shunt field control for the remainder of this experiment.
13. **Turn off the power supply and disconnect the circuit. Replace the components in their proper location.**
14. Using the measured values from the chart in Figure 28-5, draw a characteristic curve for this generator. This is done by plotting the points for output voltage and output current on the graph shown in Figure 28-6. When the points have been plotted, draw through the points with a continuous line.

Resistance (ohms)	Output voltage (volts)	Output current (amps)	Field current (amps)	AC current (amps)
Infinity				
1200				
600				
400				
300				
240				
150				
100				
80				
57.1				

FIGURE 28-5

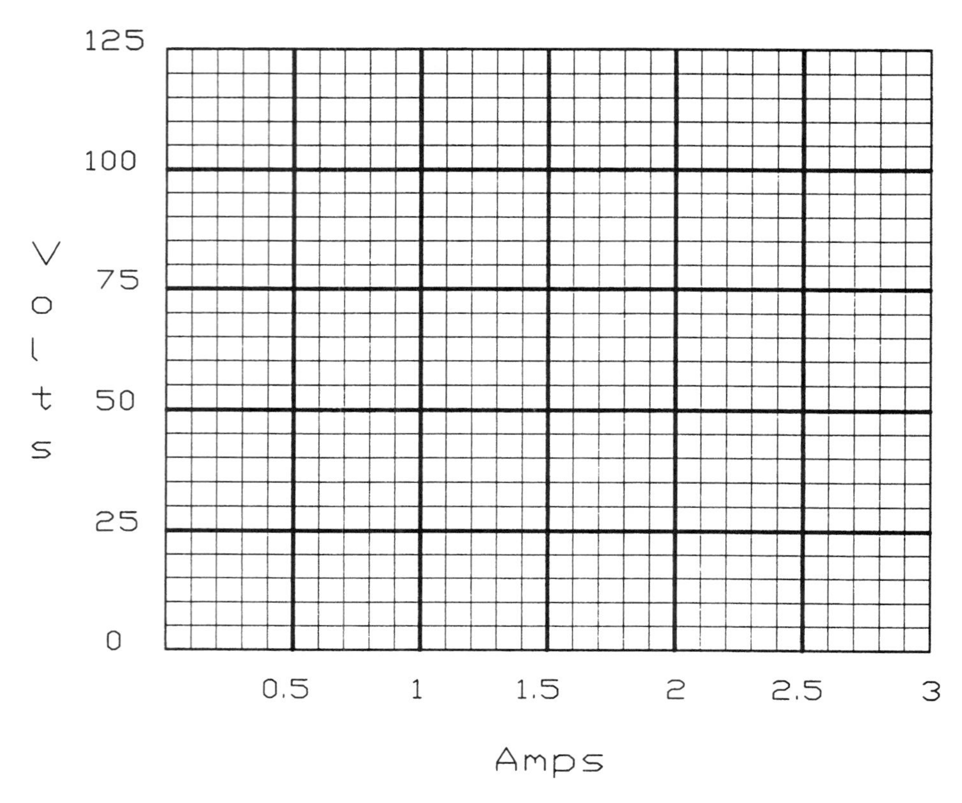

FIGURE 28-6 Grid for drawing a characteristic curve

Review Questions

1. Why must a self-excited generator have residual magnetism in its pole pieces?

____

2. How is the shunt field rheostat connected to the shunt field winding?

____

3. Why is the voltage drop of a self-excited generator greater than the voltage drop of a separately excited generator?

____

4. Why did the shunt field current decrease as load was added to the generator?

____

5. Why did the current flow to the synchronous machine increase as load on the generator increased?

____

Exercise 29

Cumulative Compound Direct Current Generators

Objectives

After completing this lab you should be able to:

- Discuss the principles of a compound generator.
- Describe over, under, and flat compounding.
- Connect a compound generator and make measurements using measuring instruments.
- Draw characteristic curves of compound generators.

Materials and Equipment

Power supply module	EMS 8821
DC metering module	EMS 8412
AC ammeter module	EMS 8425
Direct current machine module	EMS 8211
Synchronous machine module	EMS 8241
or prime mover/dynamometer module	EMS 8960
Variable-resistance module	EMS 8311
Three-phase rheostat module	EMS 8731
Ohmmeter (supplied by student)	

Discussion

The cumulative compound generator is the most used direct current generator in industry. The word *cumulative* indicates that the series and shunt fields are connected in such a manner that they aid each other in the production of magnetism. For example, assume that current flowing through the shunt field produces a south magnetic field at the pole piece. If the generator is connected as a cumulative compound, the series field is connected in such a manner that current flow through it will produce a south magnetic field on the same pole piece.

The word *compound* means that the generator has both the series and shunt fields connected. There are two basic types of cumulative compound connections:

1. the long shunt compound and
2. the short shunt compound.

The long shunt compound connection has the shunt field connected in parallel with both the armature and series field, as in Figure 29-1. The short shunt compound connection has the shunt field connected in parallel with the armature, but in series with the series field, as in Figure 29-2.

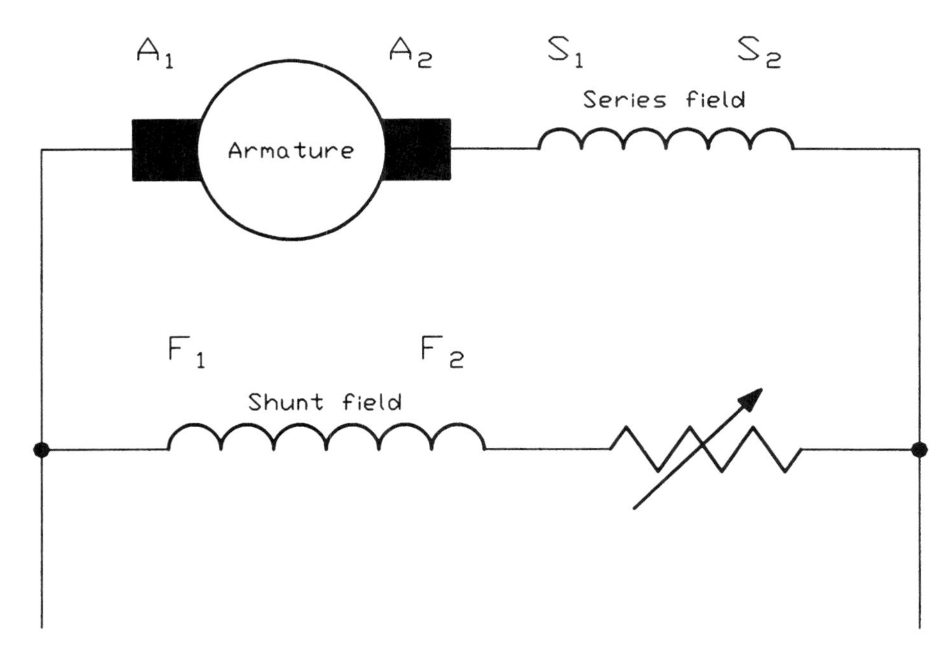

FIGURE 29-1 Long shunt connection

Both the long shunt and short shunt connections are used through-out industry. They have similar operating characteristics. The short shunt connection will sometimes produce a greater increase in voltage at light load because the shunt field current is added to the armature current as it flows through the series field. At light load, the shunt field current furnishes a higher percentage of the total amount of current flow

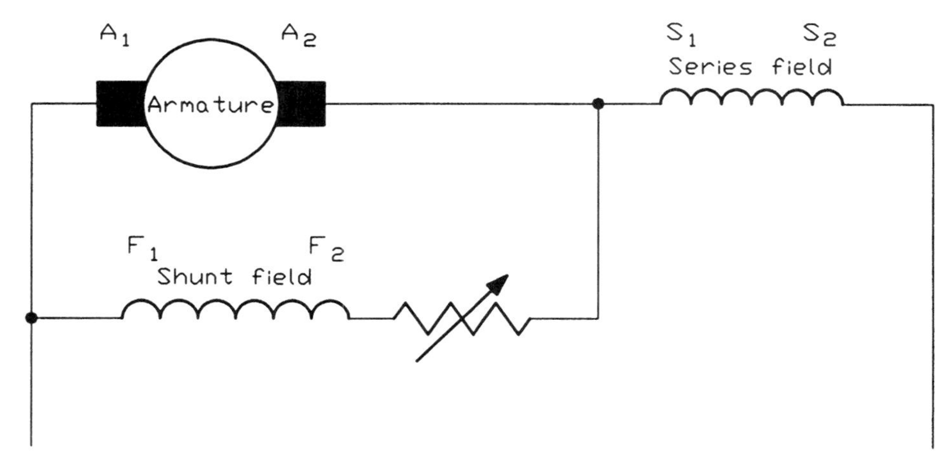

FIGURE 29-2 Short shunt connection

through the series field. The shunt field current is generally a very small percentage of the total current, as compared with the armature current, when the machine reaches full load, however, and the effect at full load is slight. It is normal for the output voltage to increase as load is first added, but it should drop back to its original value as the generator approaches full load.

If the series field is too strong, it can produce a condition known as overcompounding. Overcompounding is characterized by the fact that the output voltage at full load is higher than the output voltage at no load. Most direct current generators can be overcompounded when connected as long shunt or short shunt.

When a generator has the same output voltage at full load as it does at no load, it is flat-compounded. Flat compounding can be accomplished by adjusting the strength of the series field so that it increases the output voltage only enough to overcome the voltage losses of the machine. The most common method of controlling compounding is with the connection of a series field shunt rheostat. The series field shunt rheostat is connected in parallel with the series field as shown in Figure 29-3.

The series field shunt rheostat generally has a low value of resistance, 10 Ω or less. Since the rheostat is connected in parallel with the series field, part of the current that would normally flow through the series field will now flow through the rheostat. The reduced current flow through the series field produces less magnetic field strength in the pole pieces and, therefore, less output voltage.

If the series field is weakened too much, however, the generator will become undercompounded. Undercompounding is characterized by the fact that the output voltage at full load is less than the output voltage at no load. In this condition, the series field will not strengthen the magnetic field enough when load is added, and the voltage will drop.

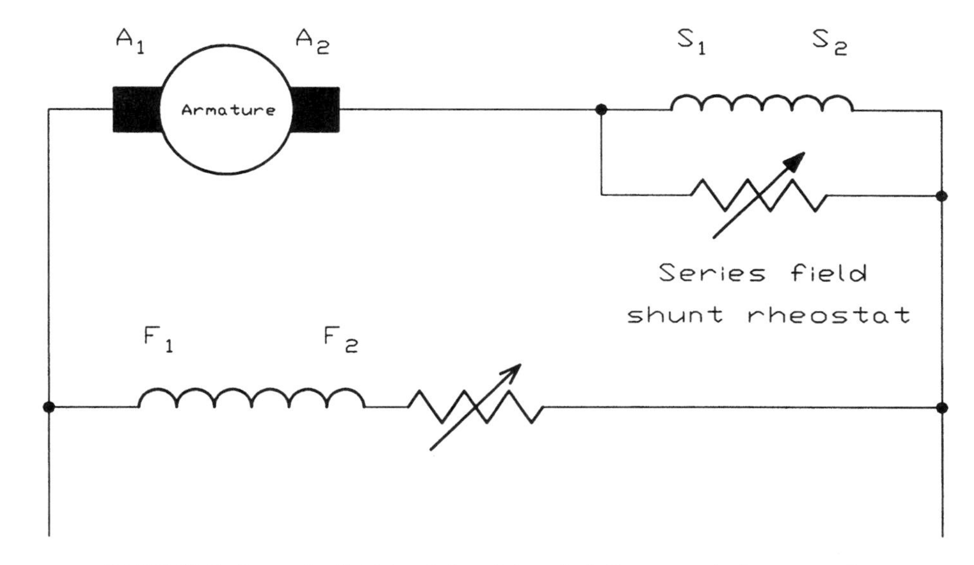

FIGURE 29-3 The series field shunt rheostat controls the amount of compounding.

Procedure

1. Open the face plate of the direct current machine and the synchronous machine or the prime mover/dynamometer, depending on which machine is being used. The prime mover/dynamometer may be employed in place of the synchronous machine.
2. Connect the two machines together with the timing belt. Be sure the timing belt is between the spring-loaded bearings.
3. Connect the circuit shown in Figure 29-4 if the synchronous machine is being used as the prime mover in this experiment. If the prime mover/dynamometer is being used, connect the circuit shown in Figure 29-5.
4. If the synchronous machine module is being used, open (turn off) switch S1 on the synchronous machine's module. Open all the switches on the variable resistance module. The synchronous machine will provide a constant speed of 1800 RPM (revolutions per minute) and is used to supply the turning force for the DC machine.

   CAUTION: *The synchronous machine should never be started with switch S1 in the closed or on position. Switch S1 should be closed only after the motor is running.*

   If the EMS 8960 prime mover/dynamometer is being used instead of the synchronous machine, it should be set for a speed of 1800 RPM and maintained throughout the experiment. Make certain the MODE switch is set in the prime mover position and the DISPLAY switch is set in the speed position.
5. Momentarily turn on the power and observe the direction of rotation of the synchronous machine and DC generator. These machines should turn in the clockwise direction as you face them. If they do not, switch the wires connected to terminals 1 and 2 of the synchronous machine. This will cause the direction of rotation to reverse.
6. Connect the circuit shown in Figure 29-6.

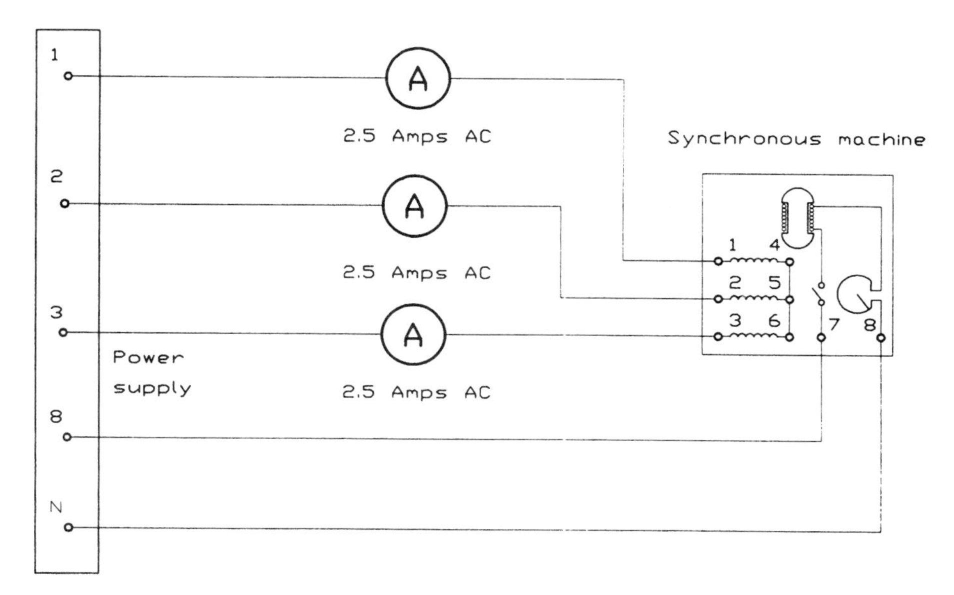

FIGURE 29-4 Synchronous motor connection

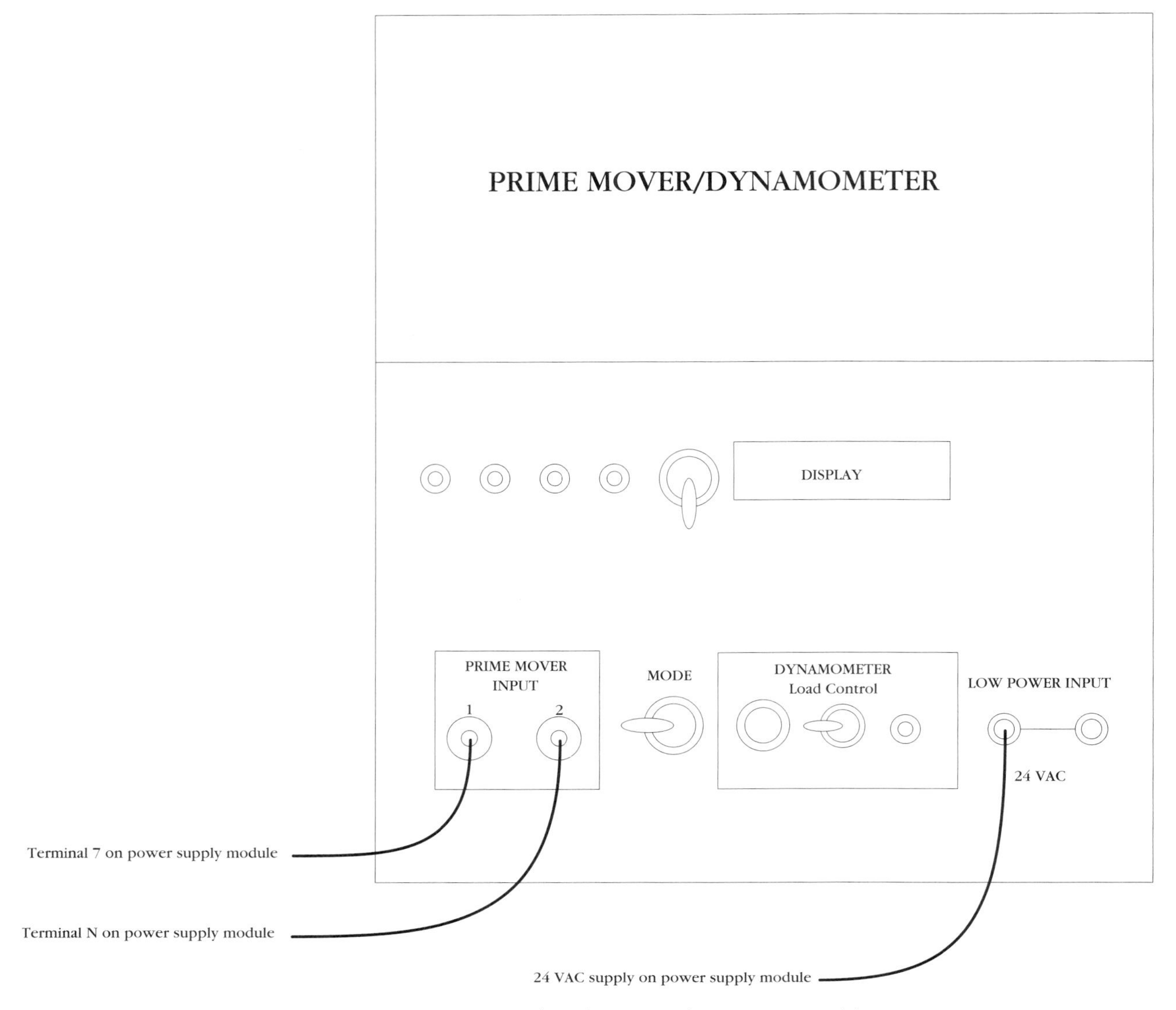

FIGURE 29-5 Connecting the prime mover/dynamometer module.

7. Turn on the power supply and close switch S_1 on the synchronous machine module, or set the prime mover for 1800 RPM
8. If the synchronous machine is being used, adjust the control knob on the front of the synchronous machine module until the three AC ammeters indicate the lowest value of current.
9. Adjust the shunt field rheostat control knob located on the front of the direct current machine module until the generator has an output voltage of 120 VDC.
10. Fill in the chart shown in Figure 29-7 by changing the resistance values on the variable-resistance module.
11. Open all the switches on the variable-resistance module, return the shunt field rheostat control knob to the full counterclockwise position, **turn off the power supply, and open switch S_1 on the synchronous machine module.**
12. Does the generator appear to be over-, flat-, or undercompounded?

____

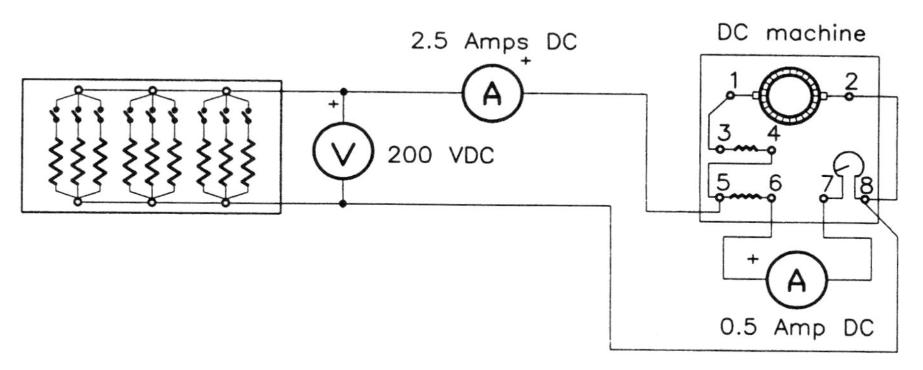

FIGURE 29-6 Connecting the long shunt compound machine

13. Reconnect the circuit as shown in Figure 29-8. Notice that one of the resistors in the three-phase rheostat module is being used as a series field shunt rheostat. Terminals 1 and N of the module are connected in parallel with the series field of the direct current machine. The three-phase rheostat module is shown in Figure 29-9
14. Disconnect the leads 1 and N on the three-phase rheostat module.
15. Connect an ohmmeter to terminals 1 and N on the three-phase rheostat module and adjust the control knob for a resistance of 6.5 Ω.
16. Disconnect the ohmmeter leads from the rheostat and reconnect the leads to the series field.
17. Turn on the power and close switch S_1 on the synchronous machine module, or set the prime mover for 1800 R.P.M.
18. Adjust the shunt field rheostat control until the generator has an output voltage of 120 V DC.
19. Fill in the chart shown in Figure 29-10 by changing the resistance values of the variable-resistance module.
20. Does this generator appear to be over-, flat-, or undercompounded?

____

Resistance (ohms)	Output voltage (volts)	Output current (amps)	Field current (amps)	AC current (amps)
Infinity				
1200				
600				
400				
300				
240				
150				
100				
80				
66.7				

FIGURE 29-7

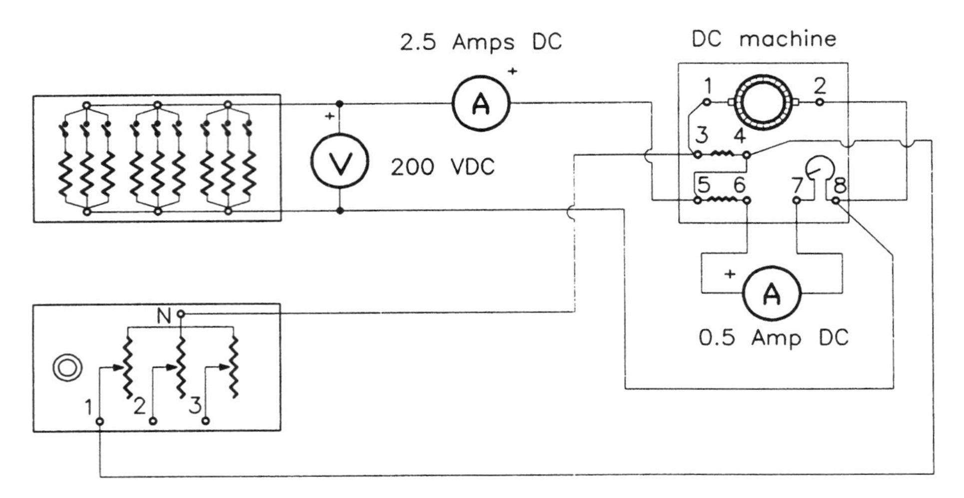

FIGURE 29-8 Controlling the amount of compounding

21. Open all the switches on the resistance module, return the shunt field rheostat control knob to the full counterclockwise direction **turn off the power supply, and open switch S_1 on the synchronous machine.**
22. Use the values obtained in the chart shown in Figure 29-7 to plot a characteristic curve on the graph shown in Figure 29-11.
23. Use the values obtained in the chart shown in Figure 29-10 to plot a characteristic curve on the graph shown in Figure 29-11.

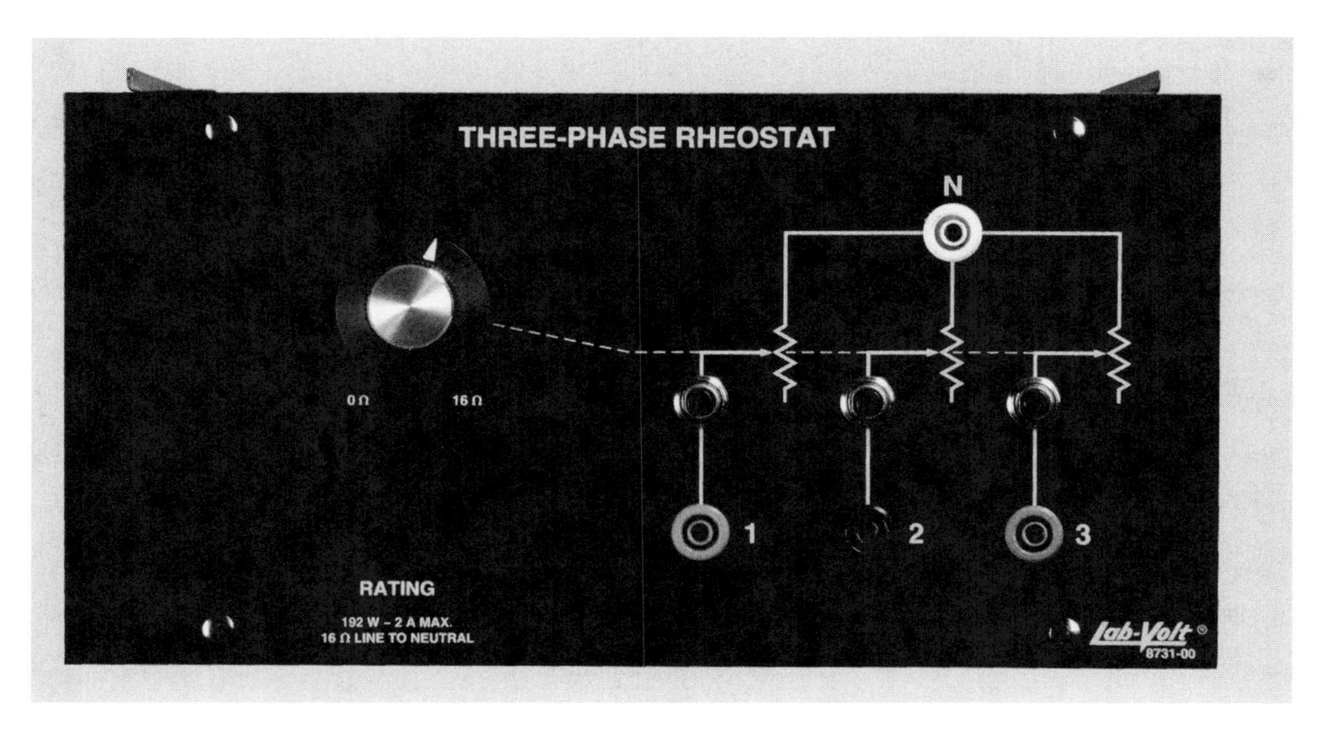

FIGURE 29-9 Three-phase rheostat module. (Courtesy of Lab-Volt® Systems, Inc.)

Resistance (ohms)	Output voltage (volts)	Output current (amps)	Field current (amps)	AC current (amps)
Infinity				
1200				
600				
400				
300				
240				
150				
100				
80				
66.7				

FIGURE 29-10

24. In the previous two examples, the direct current machine module was connected long shunt. In the next example, the machine will be connected short shunt. Measurements will be taken, and these will be compared with the previous measurements. A characteristic curve will then be plotted for comparison with the other two.

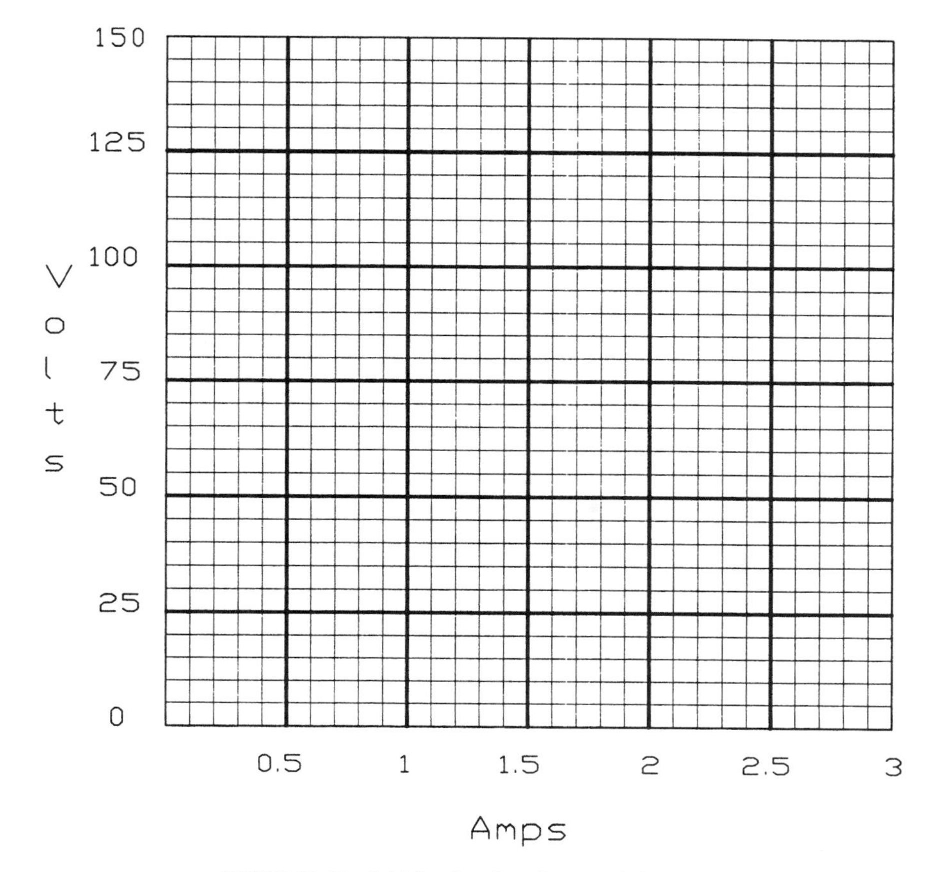

FIGURE 29-11 Grid for drawing characteristic curves

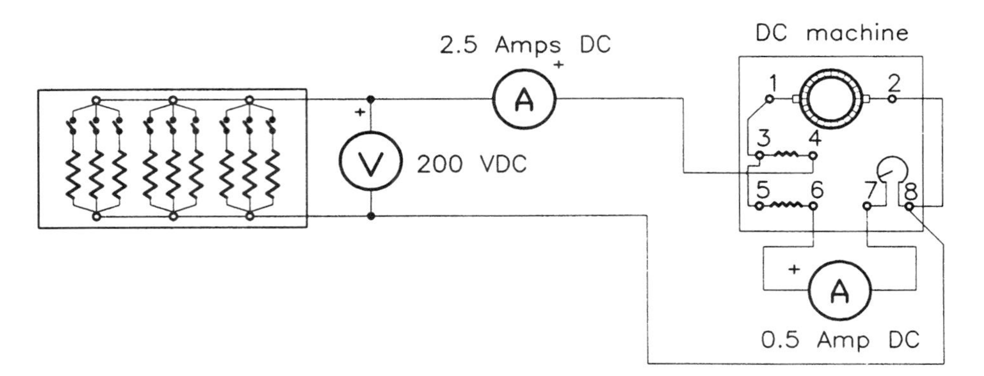

FIGURE 29-12 Short shunt compound connection

25. Connect the circuit shown in Figure 29-12.
26. Turn on the power and close switch S_1 on the synchronous machine module, or set the prime mover for 1800 RPM
27. Adjust the shunt field rheostat on the direct current machine module to permit an output voltage of 120 VDC.
28. Fill in the chart in Figure 29-13 by closing the proper switches on the resistance load module.

Resistance (ohms)	Output voltage (volts)	Output current (amps)	Field current (amps)	AC current (amps)
Infinity				
1200				
600				
400				
300				
240				
150				
100				
80				
66.7				

FIGURE 29-13

29. Open all switches on the variable-resistance module and return the shunt field rheostat to its full counterclockwise position. **Turn off the power supply and open switch S_1 on the synchronous machine module.** Disconnect the circuit and return all components to their proper place.

30. Plot a characteristic curve of the short shunt connection on the graph in Figure 29-11 using the information obtained from the chart in Figure 29-13. Compare this curve with the previous two.

Review Questions

1. What is a compound generator?

___

2. What does the word *cumulative* mean in reference to a direct current generator?

___

3. What characterizes the following conditions?

   Overcompounded ___

   Flat-compounded ___

   Undercompounded ___

4. What causes a generator to become overcompounded?

___

5. What is generally used to control the amount of compounding in a direct current generator?

___

6. Explain the difference between a long shunt compound connection and a short shunt compound connection.

___

___

Exercise 30

Differential Compound Direct Current Generators

Objectives

After completing this lab you should be able to:

- Connect a differential compound generator.
- Discuss the operation of a differential compound generator.
- Make electrical measurements of a differential generator using measuring instruments.
- Draw a characteristic curve of a differential compound generator.

Materials and Equipment

Power supply module	EMS 8821
DC metering module	EMS 8412
AC ammeter module	EMS 8425
Direct current machine module	EMS 8211
Synchronous machine module	EMS 8241
or prime mover/dynamometer module	EMS 8960
Variable-resistance module	EMS 8311

Discussion

The differential compound generator derives its name from the fact that the series and shunt field windings are connected in such a manner that they produce different magnetic polarities. When the fields are connected in this manner, they oppose each other in the production of magnetism. For example, if the current flow through the shunt field is in such a direction that it causes the field to produce a south magnetic polarity, the current flow through the series field will cause it to produce a north magnetic polarity. Because the series field is connected in series with the armature, an increase in load current will cause the magnetic pole pieces to become weaker. A weaker magnetic field will produce less output voltage. When load is added to the generator, the output voltage of the generator will drop dramatically.

Although there are some applications for a differential compound generator, they are extremely limited. This experiment is intended to demonstrate the characteristic of a differential compound machine to acquaint you with what will happen if the machine is accidentally connected in this manner.

Name ______________________________ Date ______________

Procedure

1. Open the face plate of the direct current machine and the synchronous machine or the prime mover/dynamometer, depending on which machine is being used. The prime mover/dynamometer may be employed in place of the synchronous machine.
2. Connect the two machines together with the timing belt. Be sure the timing belt is between the spring-loaded bearings.
3. Connect the circuit shown in Figure 30-1 if the synchronous machine is being used as the prime mover in this experiment. If the prime mover/dynamometer is being used, connect the circuit shown in Figure 30-2.
4. If the synchronous machine module is being used, open (turn off) switch S1 on the synchronous machines module. Open all the switches on the variable resistance module. The synchronous machine will provide a constant speed of 1800 RPM (revolutions per minute) and is used to supply the turning force for the DC machine.

   CAUTION: *The synchronous machine should never be started with switch S1 in the closed or on position. Switch S1 should be closed only after the motor is running.*

   If the EMS 8960 prime mover/dynamometer is being used instead of the synchronous machine, it should be set for a speed of 1800 RPM and maintained throughout the experiment. Make certain the MODE switch is set in the prime mover position and the DISPLAY switch is set in the speed position.

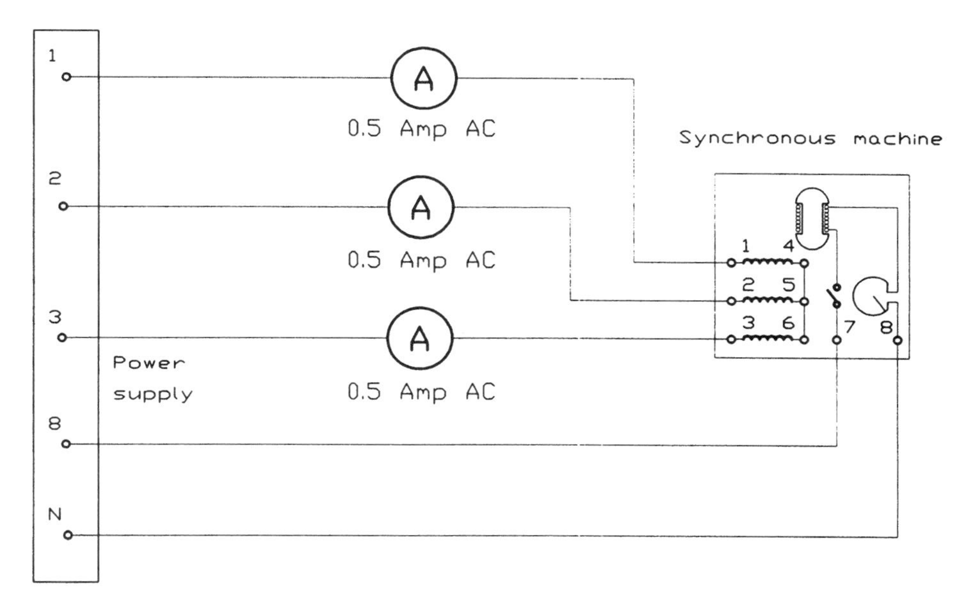

FIGURE 30-1 Synchronous motor connection

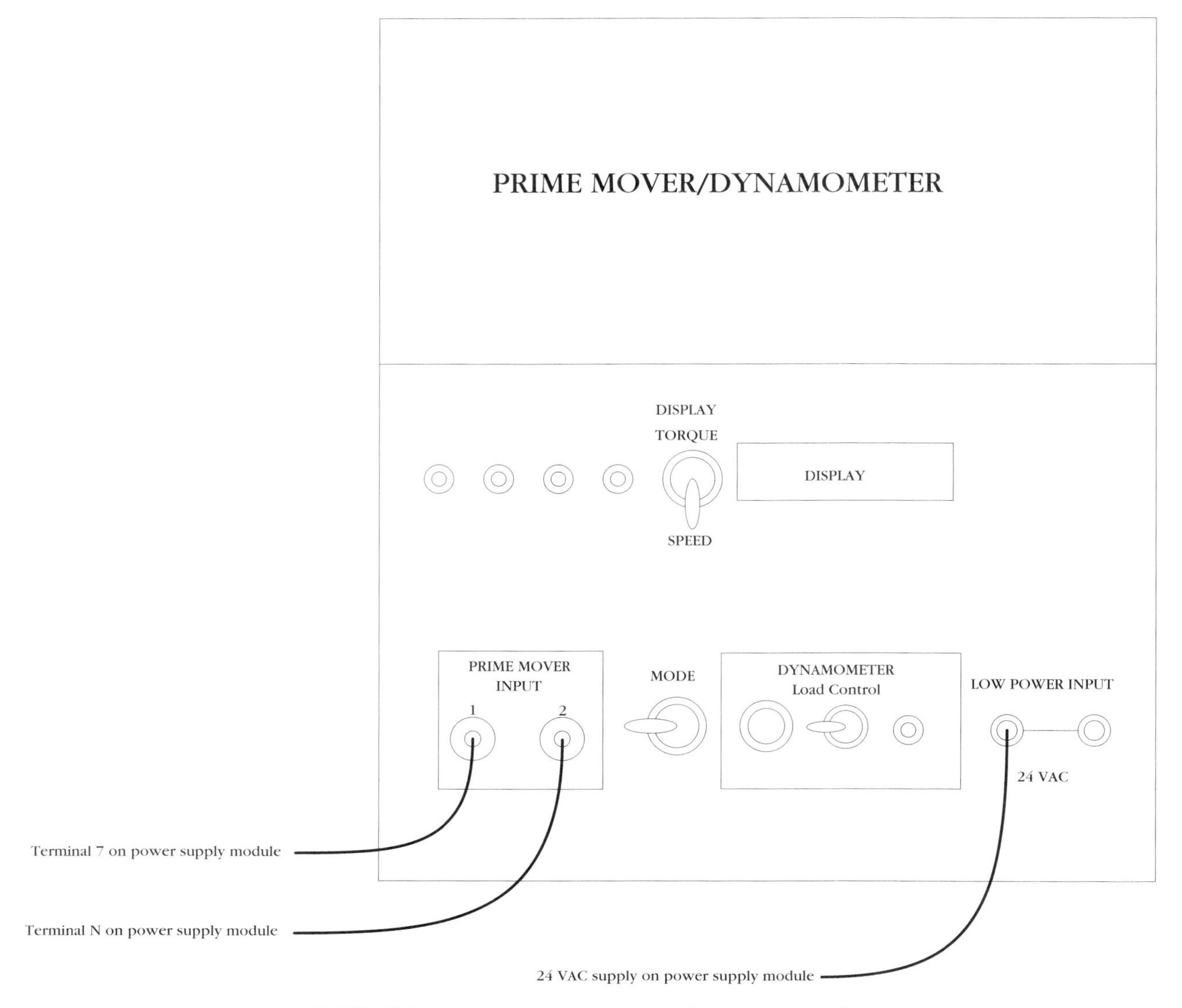

FIGURE 30-2 Connecting the prime mover/dynamometer module.

5. If the synchronous machine is being used as the prime mover, momentarily turn on the power supply and check the direction of rotation of the DC machine. It should turn in a clockwise direction as you face the machine. If it does not, **turn off the power** and reverse any two of the line leads connected to the synchronous machine.
6. Adjust the shunt field rheostat control knob on the direct current machine to the full counterclockwise position.
7. Connect the circuit shown in Figure 30-3.
8. Turn on the power supply and close switch S_1 on the synchronous machine module.

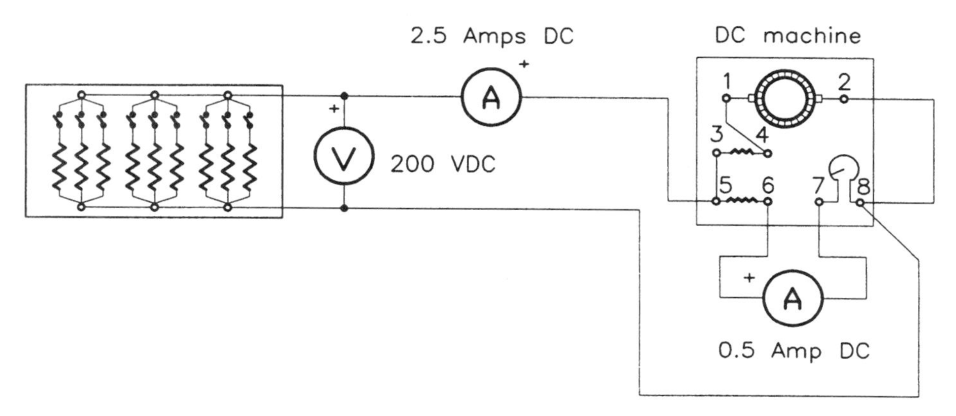

FIGURE 30-3 Differential compound generator connection

9. Adjust the control knob on the front of the synchronous machine module until the three AC ammeters indicate the lowest value of current.
10. Adjust the output voltage of the direct current generator to 120 V with the shunt field rheostat control.
11. Fill in the chart shown in Figure 30-4 by changing the resistance values on the variable-resistance module.
12. Using the information in the chart in Figure 30-4, plot a characteristic curve for the differential generator on the graph shown in Figure 30-5.
13. **Return the voltage to 0 V and turn off the power supply.**
14. Disconnect the circuit and return the components to their proper place.

Resistance (ohms)	Output voltage (volts)	Output current (amps)	Field current (amps)	AC current (amps)
Infinity				
1200				
600				
400				
300				
240				
150				
100				
80				
57.1				

FIGURE 30-4

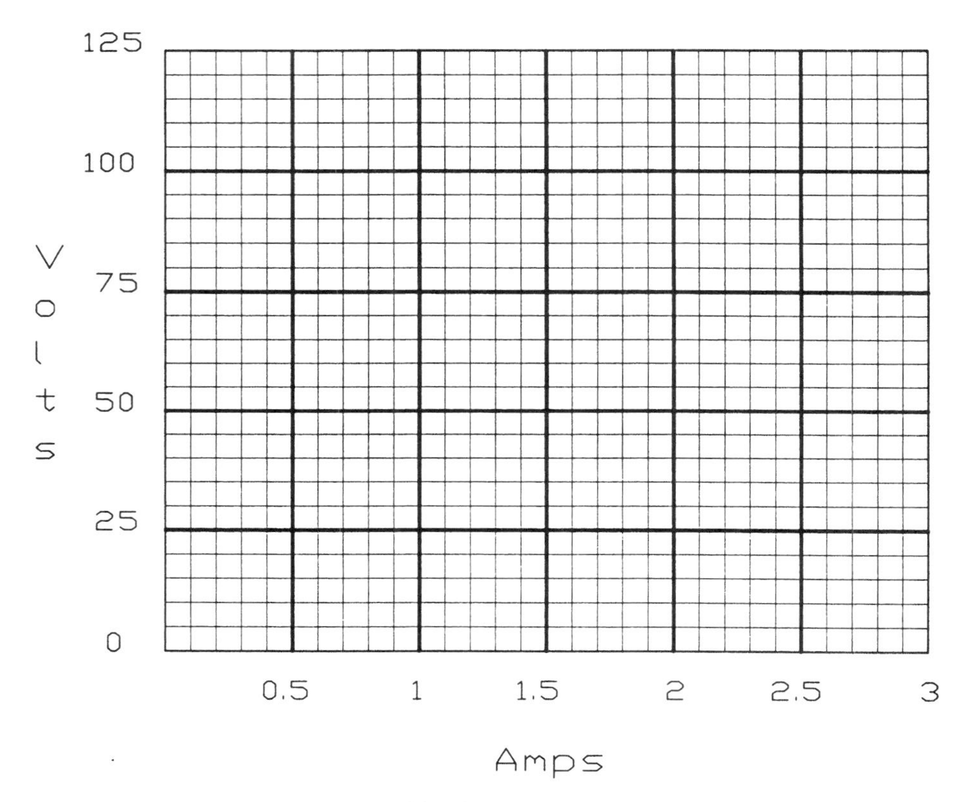

FIGURE 30-5 Grid for drawing a characteristic curve

Review Questions

1. Explain the difference between a differential compound generator and a cumulative compound generator.

____

2. If the characteristic curve of a differential compound generator is compared with that of a shunt generator, which generator has the greatest voltage drop when load is added?

____

3. Why does the differential generator have a greater voltage drop than a shunt generator?

____

Exercise 31

The Direct Current Shunt Motor

Objectives

After completing this lab you should be able to:

- Discuss the operation of a shunt motor.
- Connect a shunt motor and measure electrical quantities with measuring instruments.
- Draw a characteristic curve of a shunt motor.

Materials and Equipment

Power supply module	EMS 8821
Direct current machine module	EMS 8211
Electrodynamometer module	EMS 8911
or prime mover/dynamometer module	EMS 8960
DC metering module	EMS 8412
Hand-held tachometer	EMS 8920

Discussion

The direct current shunt motor has the shunt field connected in parallel with the armature. This arrangement permits a constant voltage source to supply current to the shunt field and maintain a constant magnetic field. The shunt motor has very good speed characteristics. The full-load speed will generally remain within 10% of the no-load speed.

The torque (turning force) produced by a direct current motor is proportional to two factors:

1. the strength of the magnetic field in the pole pieces and
2. the strength of the magnetic field in the armature.

The magnetic field strength is proportional to the amount of current flow. If a constant flow of current is maintained through the field, the magnetic strength of the pole pieces will remain constant. The amount of torque produced by the shunt motor is, therefore, proportional to the armature current.

When the windings of the armature spin through the magnetic field produced by the pole pieces, a voltage is induced into the armature. This induced voltage is opposite in polarity to the applied voltage and is known as counter-EMF (CEMF) or back-EMF. It is the counter-EMF that limits the flow of current through the armature when the motor is in operation. The amount of counter-EMF produced in the armature is proportional to three factors:

1. the number of turns of wire in the armature,
2. the strength of the magnetic field in the pole pieces, and
3. the speed of the armature.

Notice that these are the same factors that determine the amount of voltage produced by a direct current generator.

When a DC motor is first started, the inrush of current can be high because no CEMF is being produced by the armature. The only current-limiting factor is the amount of wire resistance in the windings of the armature. When current flows through the armature, a magnetic field is produced and the armature begins to turn. As the armature windings cut through the magnetic field of the pole pieces, counter-EMF opposes the applied voltage, and current flow decreases. The armature will continue to increase in speed until the counter-EMF is almost the same value as the applied voltage. At this point, the motor produces only enough torque to overcome its own losses. Some of these losses are:

1. I^2R loss in the armature windings
2. Windage loss
3. Bearing friction
4. Brush friction

When a load is added to the motor, the torque will not be sufficient to support the load at the speed the armature is turning. The armature will, therefore, slow down. When the armature slows down, counter-EMF is reduced and more current flows through the armature windings. This produces an increase in magnetic field strength and an increase in torque. This is the reason that armature current increases when load is added to the motor.

When the full rated voltage is applied to both the armature and shunt field, the motor will operate at normal, or base, speed. If full voltage is applied to the shunt field and reduced voltage is applied to the armature, the motor will operate below normal speed. If full voltage is applied to the armature and reduced voltage is applied to the shunt field, the motor will operate above normal speed.

Large DC motors are generally operated with full voltage applied to the shunt field at all times and variable voltage applied to the armature. Because voltage is applied to the shunt field at all times, the shunt field also acts as a heater to prevent formation of moisture inside the motor.

Shunt motors can be overspeeded by adding resistance in series with the shunt field. The shunt field rheostat is generally used to perform this job. When current flow is reduced in the shunt field, it causes less counter-EMF to be produced in the armature. This causes an increase of armature current and a corresponding increase of the magnetic field strength. Although the reduction of shunt field current causes a reduction in the magnetic field strength of the pole pieces, the increase of armature field strength causes a net gain in torque. This increase of torque causes an increase of speed.

The direction of rotation of the shunt motor can be reversed by changing either the shunt field leads or the armature leads but not both. Standard practice is to reverse the armature leads. The reason for this practice is that if the shunt field leads of a compound motor are reversed, the motor will change from a cumulative to a differential compound motor.

If the terminal leads of a direct current motor are known, the direction of rotation can be determined when the motor is connected to the line. Figure 31-1 shows the connection of a shunt motor to obtain different directions of rotation. The direction of rotation is determined by facing the commutator end of the armature. This is generally but not always the rear of the machine. The EMS 8211 direct current machine module is a good example of this exception.

THE DYNAMOMETER

In this experiment a dynamometer will be used to measure the amount of torque produced by the motor. There are two different types of dynamometers available, the EMS 8911 electrodynamometer and the EMS 8960 prime mover/dynamometer. The procedure in this experiment will depend on which machine you use.

The Electrodynamometer

The electrodynamometer is basically an AC induction motor that will have direct current applied to its windings to produce a braking action. The motor housing is suspended by bearings and is free to rotate. A calibrated spring is connected to the housing to produce a reverse torque as the motor turns. The front housing has a scale attached to it. The scale is calibrated to indicate pound-inches of torque and is read at the bottom of the housing. A red line is used as the indicator. An autotransformer, located on the face plate, provides variable DC voltage to the motor housing. The only electrical connection needed by the electrodynamometer is 120 VAC supplied to the two input jacks located on the front panel of the module.

Prime Mover/Dynamometer

If the prime mover/dynamometer module is being used, the only electrical connection to be made is the low-voltage 24 VAC supply that connects to the power supply module. Make sure the DISPLAY switch is set in the TORQUE position, the MODE switch is set in the DYN. position, and the MODE switch (located inside the DYNAMOMETER LOAD CONTROL box) is set in the MAN. position. The torque will be shown on the digital display in pound-inches.

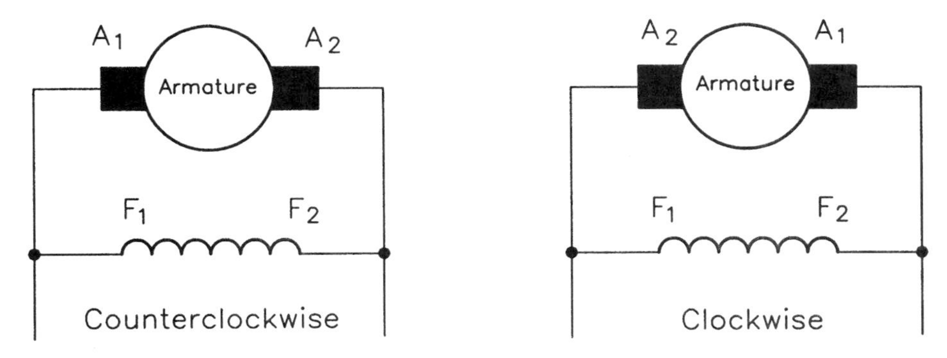

FIGURE 31-1 Determining the direction of rotation for a shunt motor

Name ______________________________ Date ______________

Procedure

1. If the electrodynamometer is being used, connect the circuit shown in Figure 31-2. If the prime mover/dynamometer is being used, connect the circuit shown in Figure 31-3.
2. Adjust the following controls to the full counterclockwise position:
   A. The voltage control knob on the power supply.
   B. The shunt field rheostat located on the direct current machine module.
   C. The voltage control knob located on the front of the electrodynamometer if it is being used.
   D. The load control knob on the front of the prime mover/dynamometer module if it is being used.
3. Turn on the power supply and adjust the shunt field rheostat until a current of 400 mA (0.4 A) flows through the shunt field.
4. Increase the voltage control knob located on the power supply until a voltage of 120 V is applied to the armature. The armature should be turning in a clockwise direction. If it is not, return the voltage to 0 V, change leads 1 and 2 on the direct current machine module, and return the voltage to 120 V. *The 120-V setting must be maintained throughout the experiment.* It may be necessary to adjust the voltage periodically.
5. Use the hand-held tachometer to adjust the shunt field rheostat located on the direct current machine module until the motor operates at a speed of 1800 RPM.

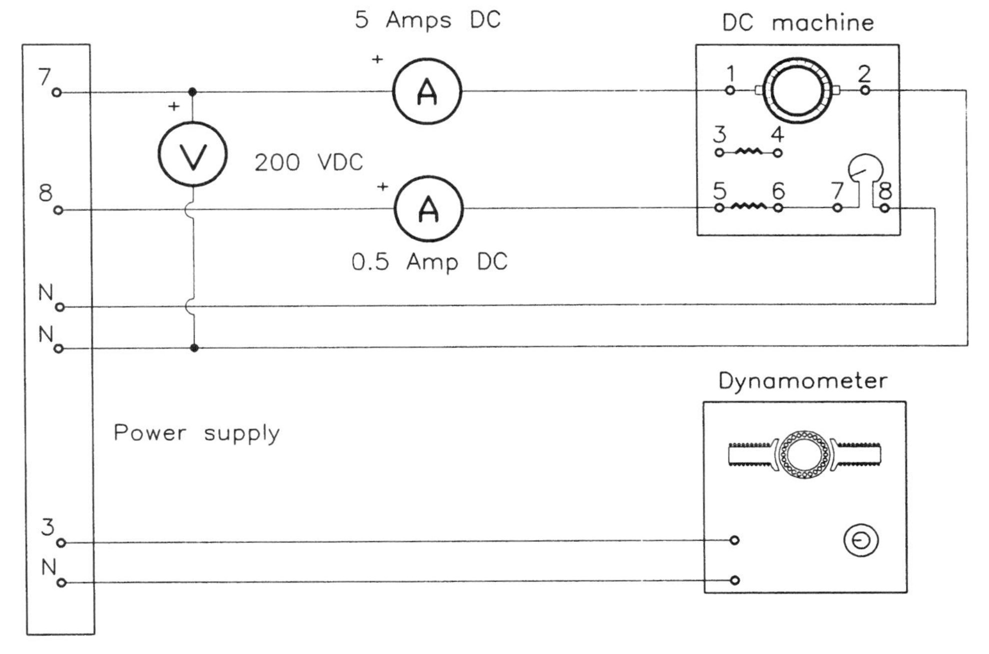

FIGURE 31-2 Shunt motor connection

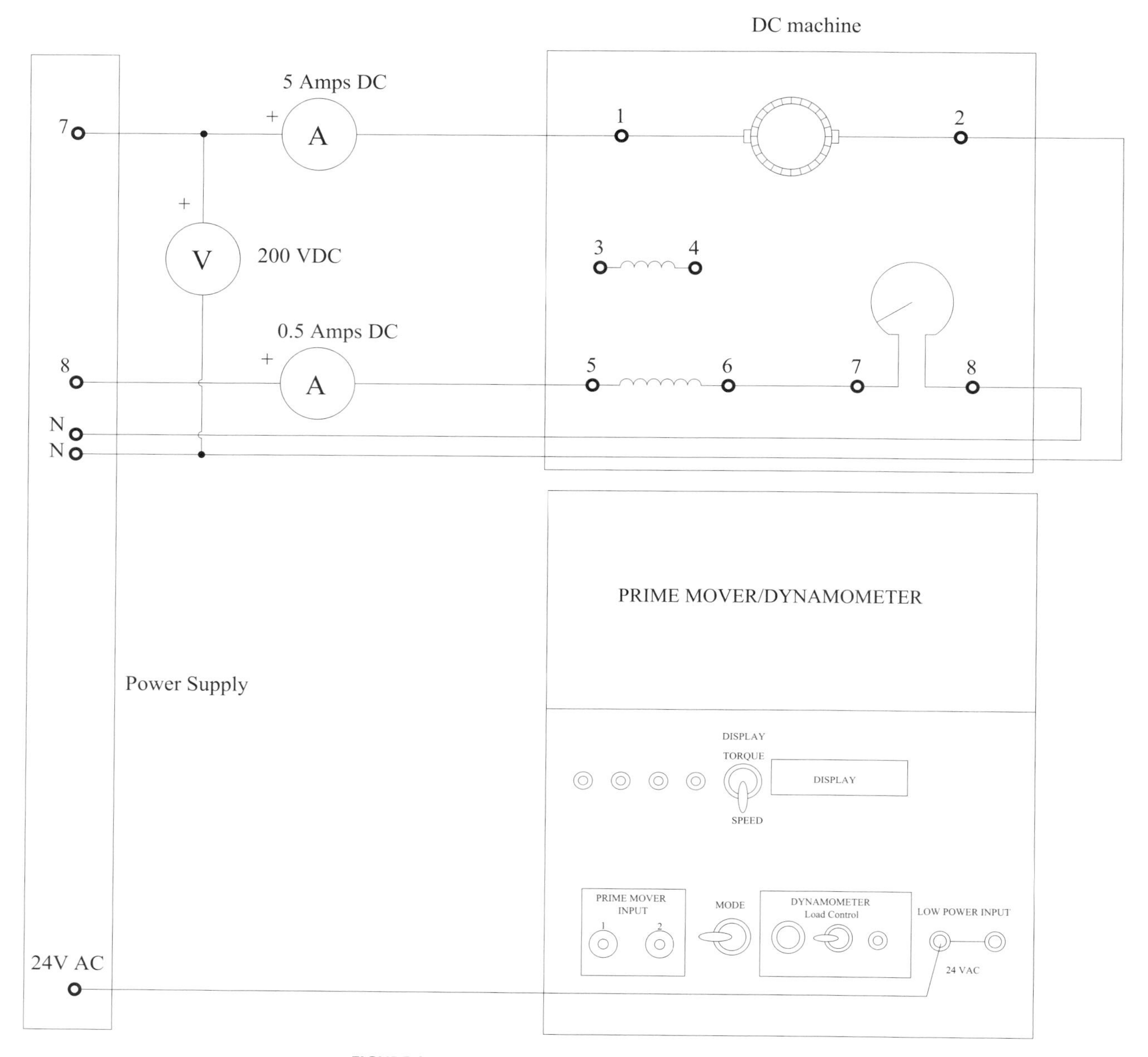

FIGURE 31-3 Prime mover/dynamometer connection

6. Fill in the values in the chart shown in Figure 31-4 by increasing the torque settings with the electrodynamometer.

NOTE: The first measurement will be taken with the timing belt disconnected between the motor and electrodynamometer. This will produce a more accurate measurement for the no-load setting.

The output horsepower value can be found using the formula

$$hp = \frac{(1.59)(\text{torq.})(\text{RPM})}{100,000}$$

where hp = horsepower
1.59 is a constant

Torq. = torque in lb-in.
RPM = speed in revolutions per minute
100,000 is a constant

Find the horsepower of a motor producing 9 lb-in. of torque at a speed of 1750 RPM.

$$hp = \frac{(1.59)(\text{torq.})(\text{RPM})}{100,000}$$
$$hp = \frac{(1.59)(9)(1750)}{100,000}$$
$$hp = \frac{25042.5}{100,000}$$
$$hp = 0.25 \text{ hp}$$

One horsepower is equal to 746 W. The output power (output watts) can be computed by multiplying the output horsepower value by 746. Assume that the motor is producing an output horsepower of 0.20 hp. The output power is:

$$746 \times 0.20 \text{ hp} = 149.2 \text{ W}$$

The input power can be computed by multiplying the voltage applied to the armature by the current flow through the armature.

$$\text{input watts} = \text{armature voltage} \times \text{armature current}$$

NOTE: This formula does not include the power consumption of the shunt field. This formula will permit the efficiency of the armature circuit to be computed. If the efficiency of the entire motor is to be computed, the power consumed by the shunt field must be added to the input power of the armature.

The efficiency can be computed using the formula

$$\text{eff.} = \frac{\text{power out}}{\text{power in}} \times 100$$

Load torque (lb-in.)	Speed (RPM)	Armature current (amps)	Output Horsepower (hp)	Output power (watts)	Input power (watts)	Eff. (%)
0						
2						
4						
6						
8						
10						
12						

FIGURE 31-4

Assume that the output power of a motor has been computed at 210 W. Also assume that the armature current is 2.5 A and that the armature voltage is 120 V. To find the efficiency of the motor, first find the input power:

$$\text{Input power} = 2.5\ \text{A} \times 120\ \text{V}$$
$$\text{Input power} = 300\ \text{W}$$

Next, compute the efficiency:

$$\text{eff.} = \frac{210\ \text{W}}{300\ \text{W}} \times 100$$
$$\text{eff.} = 70\%$$

7. **Turn off the power supply** and connect the motor and electrodynamometer together with the timing belt. Set the control knob on the electrodynamometer to the full counterclockwise position (0). Turn on the power and fill in the remainder of the chart (2 to 12 lb-in.) by adjusting the control knob located on the electrodynamometer. The value of load torque is indicated by the red line and scale markings seen at the bottom of the electrodynamometer. The scale values are marked in pound-inches.
8. Draw a characteristic curve of the shunt motor using the values in the chart in Figure 31-4 and the graph shown in Figure 31-5.
9. Return the dynamometer voltage control, the shunt field rheostat, and the DC voltage control to the full counterclockwise position.
10. **Turn off the power supply** and interchange armature leads 1 and 2 on the direct current machine module.

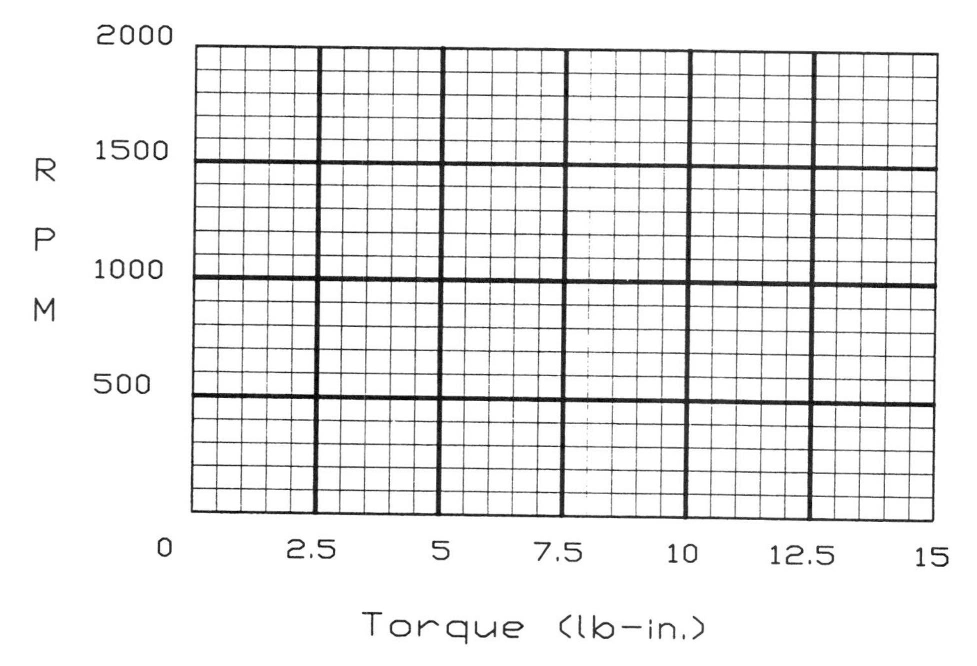

FIGURE 31-5 Grid for drawing a characteristic curve

11. Turn on the power supply and adjust the shunt field current for a value of 400 mA.
12. Slowly increase the DC voltage applied to the armature and note the direction of rotation. Has the direction of rotation changed?

    ____

13. Return the voltage control knob and the shunt field rheostat control to the full counterclockwise position and **turn off the power supply.**
14. Disconnect the circuit and return the components to their proper place.

Review Questions

1. What two factors determine the amount of torque produced by a shunt motor?

   A. ______________________________

   B. ______________________________

2. Name three factors that determine the amount of counter-EMF produced by the armature.

   A. ______________________________

   B. ______________________________

   C. ______________________________

3. Explain how a shunt motor can be underspeeded.

   ____

4. Explain how a shunt motor can be overspeeded.

   ____

5. How can the direction of rotation of a shunt motor be reversed?

   ____

6. A motor is producing 24 lb-in. of torque at a speed of 1650 RPM. How much horsepower is the motor producing?

   ______________ hp

7. What is the power output of the motor in question 6?

   ______________ W

8. Assume the motor in question 6 is connected to a 250-VDC line and has a current draw of 2.5 A. What is the efficiency of this motor?

   ______________ %

9. A motor has its A_2 lead and F_1 lead connected together and its A_1 lead and F_2 lead connected together. Will the motor run in the clockwise or counterclockwise direction?

   ____

10. What part of a DC motor should be faced when determining the direction of rotation?

    ____

Exercise 32

The Direct Current Series Motor

Objectives

After completing this lab you should be able to:

- Discuss the operation of a series-connected DC motor.
- Connect a series motor and make electrical measurements using measuring instruments.
- Draw a characteristic curve of a series motor.

Materials and Equipment

Power supply module	EMS 8821
Direct current machine module	EMS 8211
Electrodynamometer module	EMS 8911
or prime mover/dynamometer module	EMS 8960
DC metering module	EMS 8412
Hand-held tachometer	EMS 8920

Discussion

The operating characteristics of the direct current series motor are very different from those of the shunt motor. The reason is that the series motor has only a series field connected in series with the armature. The armature current, therefore, flows through the series field. The speed of the series motor is controlled by the amount of load connected to the motor and the amount of motor torque. When load is increased, the speed of the motor will decrease. This decrease causes a reduction in the amount of counter-EMF produced in the armature and an increase in armature and series field current. Because the current increases in both the armature and series field, the torque will increase by the square of the current. In other words, if the current doubles, the torque will increase four times.

Series motors have no natural speed limit and should, therefore, never be operated in a no-load condition. Large series motors that suddenly lose their load will race to speeds that will destroy the motor. For this reason, series motors should be coupled directly to a load. Belts or chains should never be used to connect a series motor to a load.

Series motors have the ability to develop extremely high starting torques. An average of about 450% of full torque is common. These motors are generally used for applications that require a high starting torque, such as the starter motor on an automobile, cranes and hoists, and electric buses and street cars.

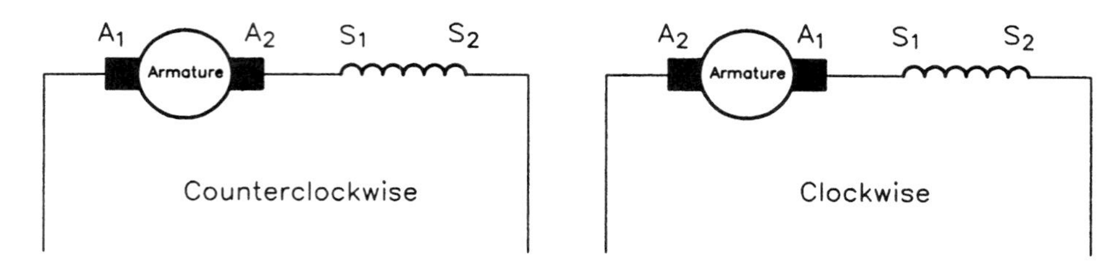

FIGURE 32-1 Determining the direction of rotation for a series motor

The direction of rotation of a series motor can be changed by reversing the armature leads or the series field leads. Figure 32-1 shows standard motor connections for both clockwise and counterclockwise rotation.

Name ______________________________ Date ______________

Procedure

1. Connect the direct current machine and the dynamometer together with the timing belt.
2. If the electrodynamometer is being used, connect the circuit shown in Figure 32-2. If the prime mover/dynamometer is being used, connect the circuit shown in Figure 32-3.
3. Adjust the control knob located on the power supply to the full counterclockwise position.
4. Adjust the control knob on the dynamometer to about 50% of full setting.
5. Turn on the power supply and increase the voltage control knob located on the power supply until a voltage of 120 V is applied to the armature. The armature should be turning in a clockwise direction as you face it. If it is not, return the voltage to 0 V, change leads 1 and 2 on the direct current machine module, and return the voltage to 120 V. The 120-V setting must be maintained throughout the experiment.
6. Adjust the dynamometer rheostat for a torque of 2 lb-in.
7. Fill in the values in the chart shown in Figure 32-4 by increasing the torque settings with the electrodynamometer.

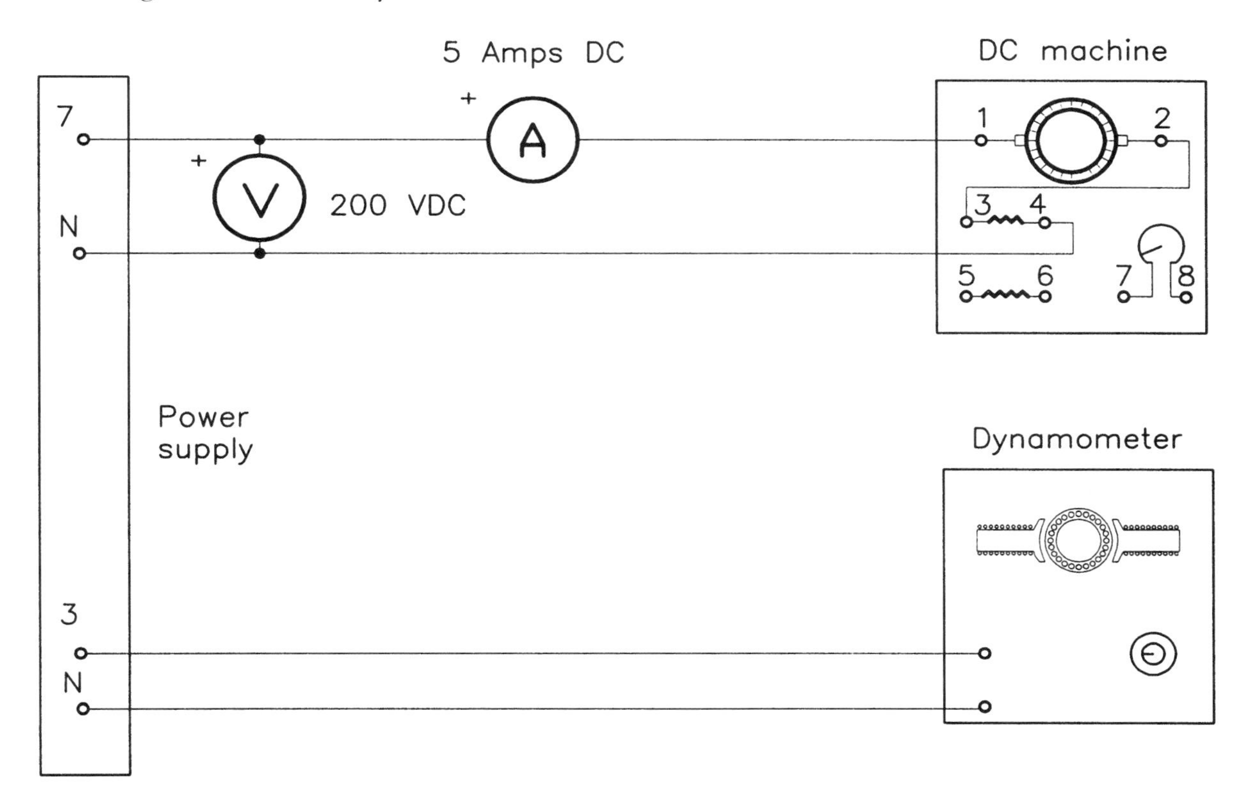

FIGURE 32-2 Series motor connection

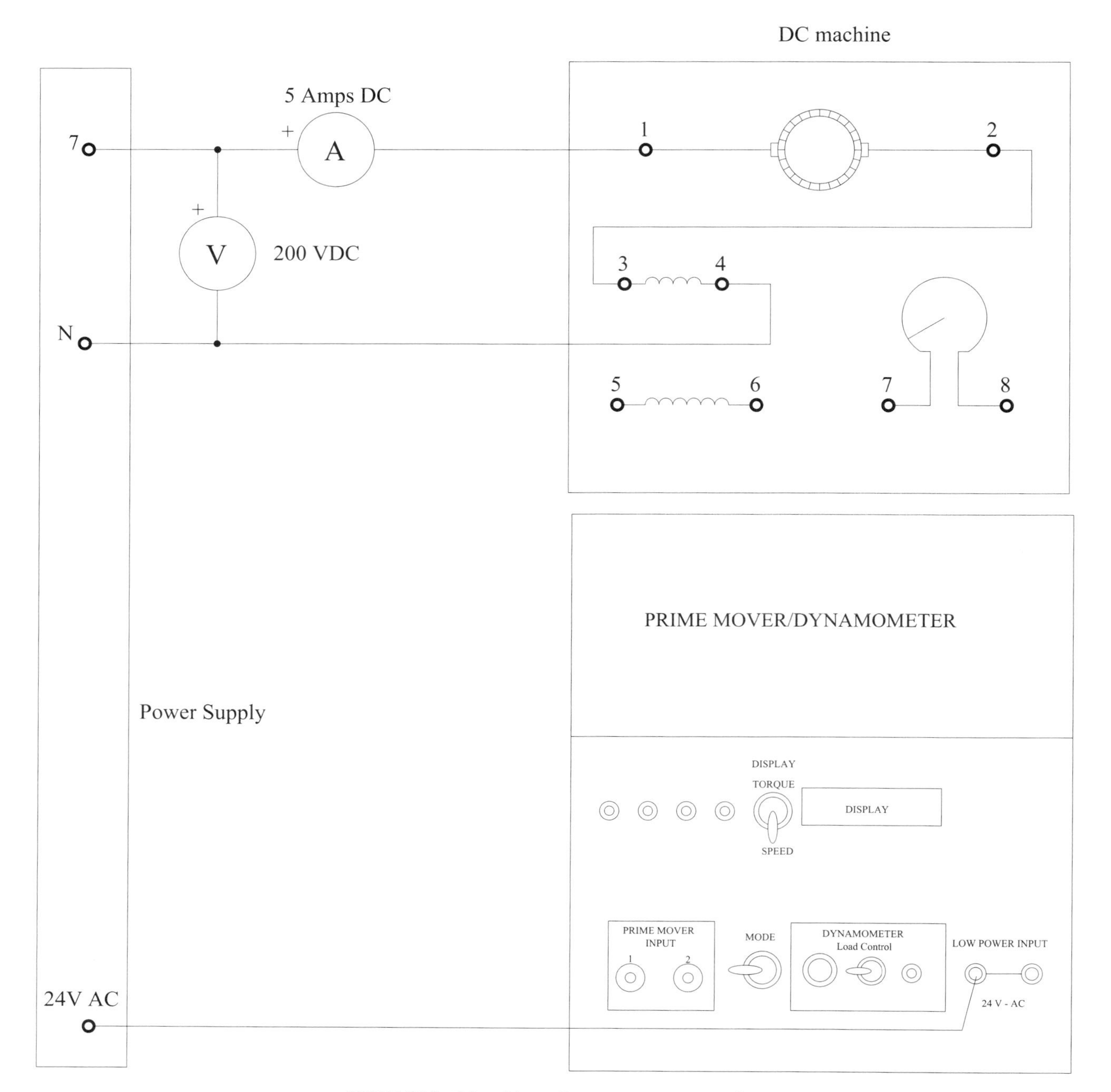

FIGURE 32-3 Prime Mover/Dyanmometer connection

The output horsepower value can be found using the formula

$$hp = \frac{(1.59)(\text{torq.})(\text{RPM})}{100,000}$$

where hp = horsepower
1.59 is a constant
Torq. = torque in lb-in.
RPM = speed in revolutions per minute
100,000 is a constant

Load torque (lb-in.)	Speed (RPM)	Armature current (amps)	Output Horsepower (hp)	Output power (watts)	Input power (watts)	Eff. (%)
0						
2						
4						
6						
8						
10						
12						

FIGURE 32-4

The output power (output watts) can be computed by multiplying the output horsepower value by 746.

The input power can be computed by multiplying the voltage applied to the armature by the current flow through the armature.

$$\text{Input watts} = \text{armature voltage} \times \text{armature current}$$

The efficiency can be computed using the formula

$$\text{eff.} = \frac{\text{power out}}{\text{power in}} \times 100$$

8. Return the voltage control knob to the full counterclockwise position and **turn off the power supply.**
9. Draw a characteristic curve of the series motor using the values in the chart in Figure 32-4 and the graph shown in Figure 32-5.
10. Readjust the dynamometer control to the full counterclockwise position.
11. Turn on the power supply and adjust the voltage to 120 V.

NOTE: The speed of the motor will be high in this experiment. However, the bearing friction and windage loss of the motor and dynamometer are used to provide enough load on the motor to limit its speed.

12. Measure the speed of the motor.

    ______________ RPM

13. Return the voltage control knob on the power supply to the full counterclockwise position and **turn off the power.**
14. Reverse the armature leads 1 and 2 on the direct current machine module.
15. Slowly increase the voltage until the motor begins to turn. Note the direction of rotation. Does the motor turn in the counterclockwise direction?

    ____

16. Return the voltage control knob to the full counterclockwise position and turn off the power supply.
17. Disconnect the circuit and return the components to their proper place.

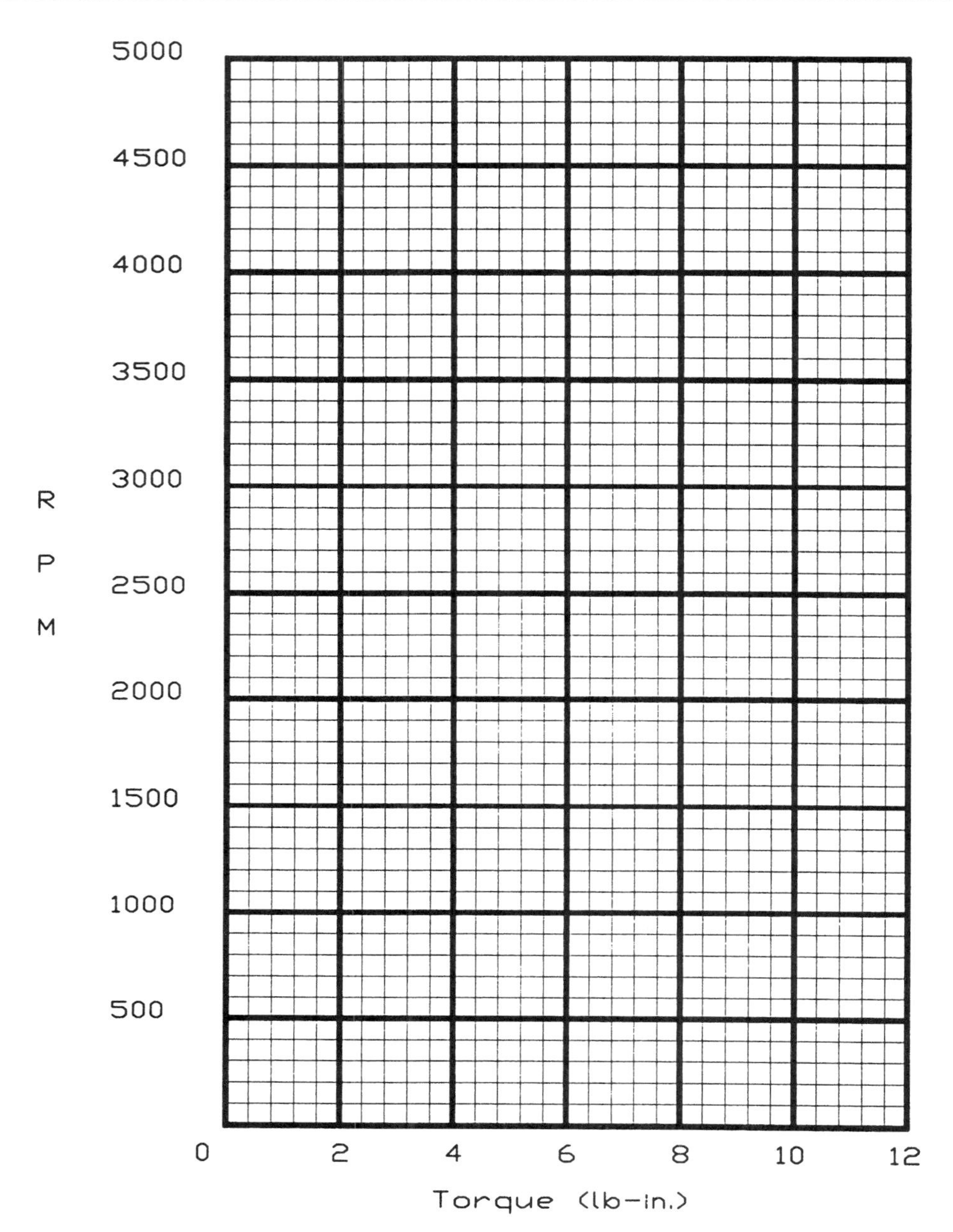

FIGURE 32-5 Grid for drawing the characteristic curve of a series motor

Review Questions

1. A series-wound DC motor is producing 3000 lb-in. of torque at a speed of 1450 RPM. The motor is connected to a 1250-VDC line and has a current draw of 63.6 A. What is the efficiency of the motor?

   ______________ %

2. How is the direction of rotation of a series motor reversed?

   ____

3. Why should a series motor never be operated without a load connected to it?

   ____

4. About what percentage of full-load torque can a series motor develop when starting?

   ______________ %

5. Does the speed of a series motor increase or decrease when load is added to it?

   ____

Exercise 33

The Direct Current Compound Motor

Objectives

After completing this lab you should be able to:

- Discuss the construction and operation of a direct current compound motor.
- Connect a direct current compound motor.
- Make electrical measurements of a compound motor using measuring instruments.
- Draw a characteristic curve of a compound motor.

Materials and Equipment

Power supply module	EMS 8821
Direct current machine module	EMS 8211
Electrodynamometer module	EMS 8911
or prime mover/dynamometer module	EMS 8960
DC metering module	EMS 8412
Hand-held tachometer	EMS 8920
Ohmmeter (supplied separately)	

Discussion

The compound motor uses both a series field and a shunt field. This motor is used to combine the operating characteristics of both the series and shunt motors. The series field of the compound motor permits the motor to develop high torque, and the shunt field permits speed control and regulation. The compound motor is used more than any other type of direct current motor in industry. The compound motor will not develop as much torque as the series motor, but it will develop more than the shunt motor. The speed regulation of a compound motor is not as good as that of a shunt motor, but it is much better than that of a series motor.

The direction of rotation is generally reversed by changing the armature leads. If only the shunt field lead were to be reversed, the motor would be changed from cumulative compound to differential compound. Care must always be taken not to let this condition occur. If a cumulative compound motor is changed to a differential compound, the motor will reverse direction and operate as a series motor when load is added. This can be extremely dangerous to both people and equipment. The direction of rotation diagrams for a cumulative compound motor are shown in Figure 33-1.

DETERMINING CUMULATIVE AND DIFFERENTIAL COMPOUND CONNECTION

The shunt field will always control the direction of rotation when the motor is operated without load or at light load, but when a heavy load is added the series field will become stronger and will determine the direction of rotation. The diagrams shown in Figure 33-1 assume that the field and armature leads have been properly marked. There may be occasions when it becomes necessary to connect a compound motor after the

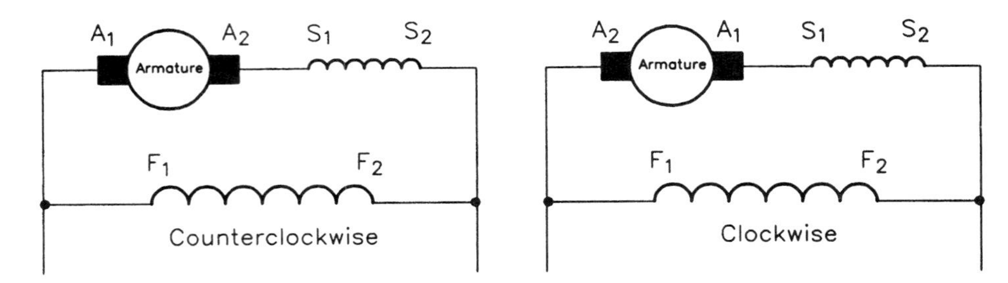

FIGURE 33-1 Determining the direction of rotation for a cumulative compound motor

lead markings have been lost or you suspect they may have been marked incorrectly. The following procedure can be used to connect the motor and ensure it is not connected differential compound.

1. Disconnect the motor from the load.
2. Use an ohmmeter to determine which leads are connected to the armature, series field, and shunt field. (This procedure was discussed in Exercise 24.)
3. Connect one of the series field leads to one of the armature leads. Connect the other armature lead to one incoming power line and the other series field lead to the other power line. The motor is now connected as a series motor.
4. Momentarily apply power to the motor just long enough to observe its direction of rotation. (The power application must be momentary. The motor is now connected as a series motor, and the load is disconnected.)
5. Connect the shunt field to form a long or short shunt connection. Again apply power to the motor and observe the direction of rotation. If the motor turns in the same direction as previously, it is connected cumulative compound. If the motor turns in the opposite direction as previously, it is connected differential compound and the shunt field leads should be reversed.
6. If the motor rotates in the opposite direction desired, reverse the armature leads.

Name ______________________________ Date ______________

Procedure

1. If the electrodynamometer is being used, connect the circuit shown in Figure 33-2. If the prime mover/dynamometer is being used, connect the circuit shown in Figure 33-3.
2. Adjust the following controls to the full counterclockwise position:
   A. The voltage control knob on the power supply.
   B. The shunt field rheostat located on the direct current machine module.
   C. The voltage control knob located on the front of the electrodynamometer if it is being used.
   D. The load control knob on the front of the prime mover/dynamometer module if it is being used.
3. Turn on the power supply and adjust the shunt field rheostat until a current of 400 mA. flows through the shunt field.
4. Increase the voltage control knob located on the power supply until a voltage of 120 V is applied to the armature. The armature should be turning in a clockwise direction. If it is not, return the voltage to 0 V, change leads 1 and 2 on the direct current machine module, and return the voltage to 120 V. *The 120-V setting must be maintained throughout the experiment.*
5. Using the hand-held tachometer, adjust the shunt field rheostat until the motor operates at a speed of 1800 RPM.
6. Fill in the values in the chart shown in Figure 33-4 by increasing the torque settings with the electrodynamometer.

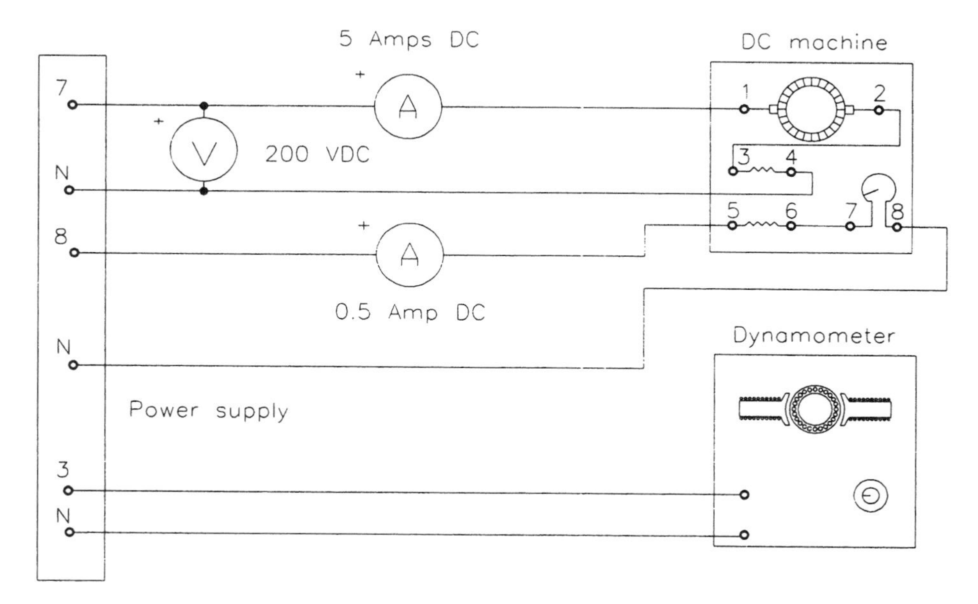

FIGURE 33-2 Cumulative compound motor connection

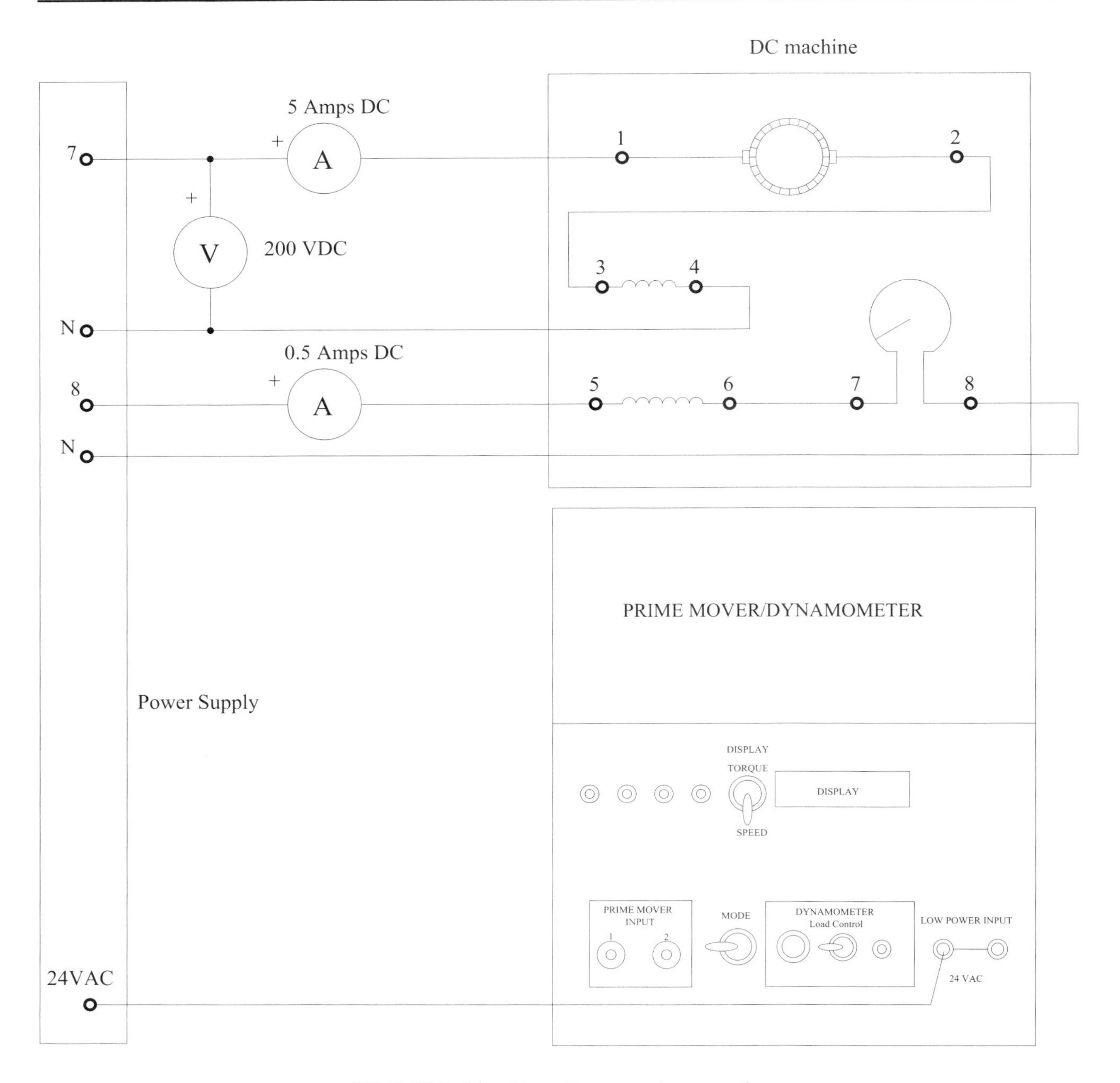

FIGURE 33-3 Prime Mover/Dynamometer connection

NOTE: The first measurement will be taken with the timing belt disconnected between the motor and electrodynamometer to provide a more accurate measurement for the no-load setting.

The output horsepower value can be found using the formula

$$\text{hp} = \frac{(1.59)(\text{torq.})(\text{RPM})}{100{,}000}$$

where hp = horsepower
1.59 is a constant

Load torque (lb-in.)	Speed (RPM)	Armature current (amps)	Output Horsepower (hp)	Output power (watts)	Input power (watts)	Eff. (%)
0						
2						
4						
6						
8						
10						
12						

FIGURE 33-4

Torq. = torque in lb-in.
RPM = speed in revolutions per minute
100,000 is a constant

The output power (output watts) can be computed by multiplying the output horsepower value by 746.

The input power can be computed by multiplying the voltage applied to the armature by the current flow through the armature.

$$\text{input watts} = \text{armature voltage} \times \text{armature current}$$

NOTE: This formula does not include the power consumption of the shunt field. This formula will permit the efficiency of the armature circuit to be computed. If the efficiency of the entire motor is to be computed, the power consumed by the shunt field must be added to the input power of the armature.

The efficiency can be computed using the formula

$$\text{eff.} = \frac{\text{power out}}{\text{power in}} \times 100$$

7. **Turn off the power supply** and connect the timing belt between the motor and dynamometer. Turn on the power supply and fill in the remainder of load values (2 to 12 lb-in) by adjusting the electrodynamometer to the proper load setting.
8. Draw a characteristic curve of the compound motor using the values in the chart in Figure 33-4 and the graph shown in Figure 33-5
9. Return the dynamometer voltage control, the shunt field rheostat, and the DC voltage control to the full counterclockwise position.
10. **Turn off the power supply** and interchange armature leads 1 and 2 on the direct current machine module.
11. Turn on the power supply and adjust the shunt field current for a value of 400 mA.
12. Slowly increase the DC voltage applied to the armature and note the direction of rotation. Has the direction of rotation changed?

____

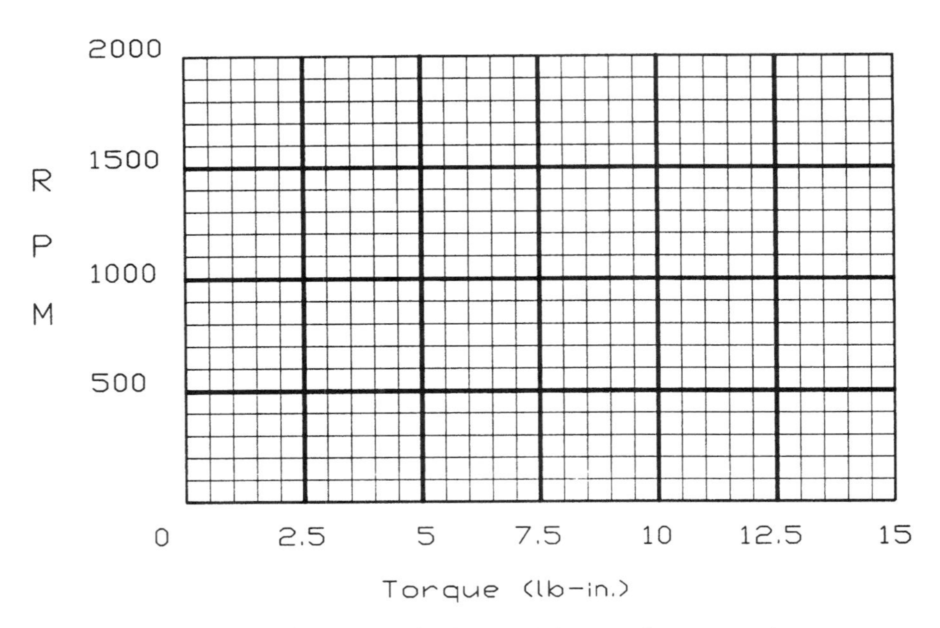

FIGURE 33-5 Grid for drawing the characteristic curve of a compound motor

13. Return the voltage control knob and the shunt field rheostat control to the full counter-clockwise position and **turn off the power supply.**
14. Disconnect the circuit and return the components to their proper place.

Review Questions

1. How is the direction of rotation of a compound motor generally reversed?

   ____

2. What would happen if the shunt field lead were changed to reverse the direction of rotation?

   ____

3. What happens to a compound motor if it is connected as a differential compound motor and load is added?

   ____

4. What is the advantage of a compound motor over a series motor?

   ____

5. What is the advantage of a compound motor over a shunt motor?

   ____

Exercise 34

The Three-phase Alternator

Objectives

After completing this lab you should be able to:

- Discuss the operation of a three-phase alternator.
- Explain the effect of speed of rotation on frequency.
- Explain the effect of field excitation on output voltage.
- Connect a three-phase alternator and make measurements using test instruments.

Materials and Equipment

Power supply module	EMS 8821
Variable-capacitance module	EMS 8331
AC ammeter module	EMS 8425
AC voltmeter module	EMS 8426
Direct current machine module	EMS 8211
or prime mover/dynamometer module	EMS 8960
Synchronous machine module	EMS 8241
DC metering module	EMS 8412
Hand-held tachometer	EMS 8920

Discussion

A three-phase alternating current generator, or alternator, is constructed by placing three sets of windings 120° apart, as in Figure 34-1. The windings are placed in a metal core around the inside of a metal housing.

An electromagnet placed in the center of the windings can be rotated around the inside of the core. As the field of the magnet cuts through the windings, a voltage is induced into the windings. The electromagnet must be excited with a source of direct

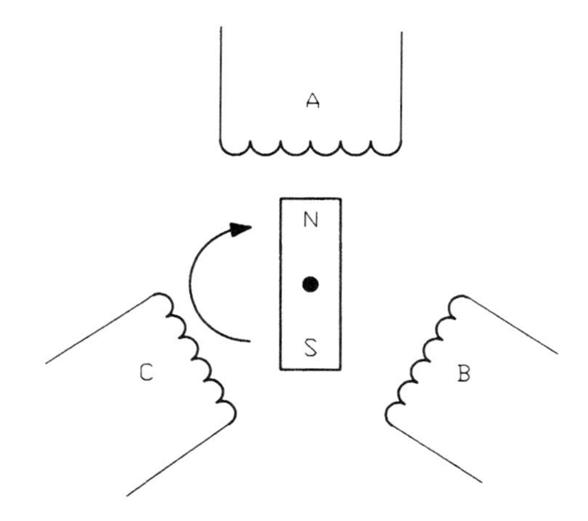

FIGURE 34-1 Three sets of windings 120° apart

current to produce the magnetic field needed. Two basic methods are employed to provide this field excitation. One method is to use a brushless exciter. A brushless exciter is constructed by adding a small set of three-phase windings to the shaft of the alternator. A set of stationary windings, similar to the pole pieces of a DC machine, provide a source of magnetic flux. The voltage induced in the three-phase winding is changed into direct current by solid state rectifiers located on the shaft. The amount of voltage produced by the brushless exciter can be controlled by varying the magnetic field strength of the stationary magnets.

The second method of providing excitation current for the rotor is to connect the winding to a set of sliprings. Sliprings are metal rings that have been electrically insulated from the shaft and each other. Brushes attached to the rings permit direct current to be applied to the rotor winding. The electromagnet winding, shaft, and sliprings, or brushless exciter, is known as the rotor, or rotating member, of the machine. The synchronous machine provided in the Lab-Volt EMS system uses sliprings and brushes to provide excitation current to the rotor.

The illustration in Figure 34-1 shows the basic concept of a three-phase alternator. In Figure 34-2, pole pieces are shown to give a more accurate illustration of how the windings are arranged. The illustration in Figure 34-2 is a two-pole alternator. Two-pole means there are two poles for each phase. Alternators with four, six, or eight poles are not uncommon. After the windings have been placed in the metal core, the entire unit is known as the stator, or stationary member, of the machine.

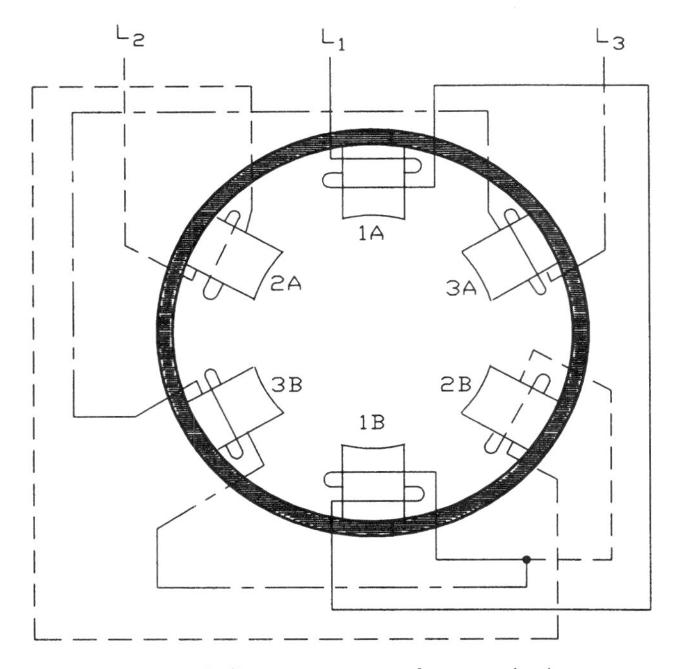

FIGURE 34-2 Winding arrangement of a two-pole alternator

The output frequency of an alternator is determined by two factors:

1. the number of stator poles and
2. the speed of rotation of the rotor.

Because the number of stator poles is constant for a particular machine, the output frequency is controlled by controlling the speed of the rotor. The chart below shows the speed of rotation needed to produce 60 Hz for alternators with different numbers of poles.

Speed (RPM)	Number of stator poles
3600	2
1800	4
1200	6
900	8

Three factors that determine the amount of output voltage of an alternator:

1. the number of turns of wire in the stator winding,
2. the strength of the magnetic field in the rotor, and
3. the speed of rotation of the rotor.

The number of turns of wire in the stator cannot be changed in a particular machine without rewinding the stator, and the speed of rotation is generally maintained at a certain level to provide a constant output frequency. Therefore, the output voltage is controlled by increasing or decreasing the strength of the magnetic field of the rotor. The magnetic field strength can be controlled by controlling the DC excitation current to the rotor.

Name ______________________________ Date ______________

Procedure

1. Remove the synchronous machine from the mobile console.
2. Locate and examine the following items on the synchronous machine module.
   A. The sliprings and brushes.
   B. The rotor winding.
   C. The stator winding.
   D. The field control rheostat located on the front panel. This rheostat is used to control the amount of excitation current applied to the rotor.
   E. The field excitation switch located on the front panel. This switch is used to apply excitation current to the rotor of the alternator.
   F. The stator winding connection located on the front panel of the synchronous machine. There are three separate windings. One set is labeled with terminal ends 1 and 4. The second set is labeled 2 and 5, and the third set is labeled 3 and 6.
3. Place the synchronous machine module in the mobile console.
4. Connect the direct current machine or the prime mover and the synchronous machine together with the timing belt. Be sure to place the timing belt between the spring-loaded bearings.
5. Connect the circuit shown in Figure 34-3 if the DC machine is being used. If the prime mover is being used, connect it as shown in Figure 34-4.
6. If the DC machine is being used as the prime mover, turn on the power supply and adjust the shunt field rheostat, located on the front of the DC machine module, until a current of 0.4 A flows through the shunt field. If the prime mover/dynamometer is being used, turn on the power supply and proceed to Step 7.

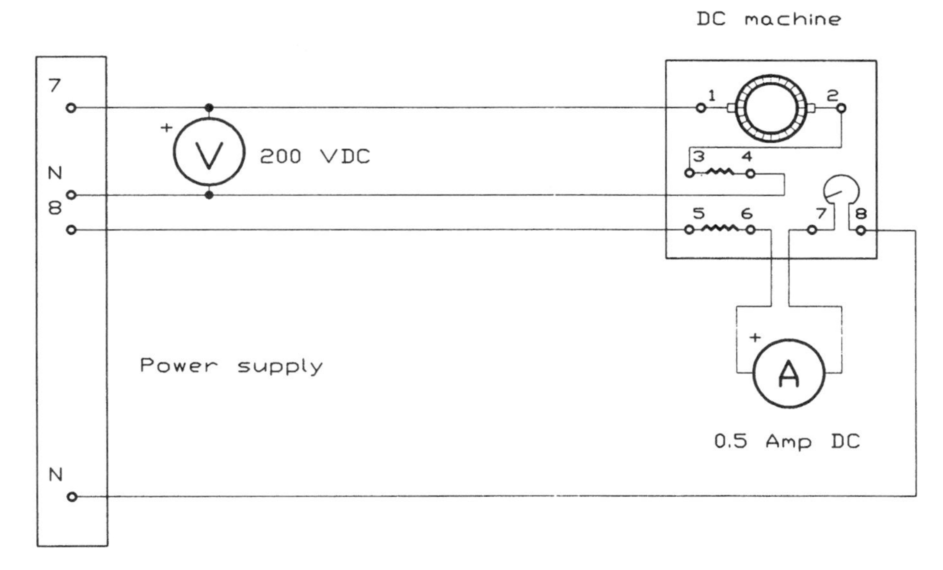

FIGURE 34-3 DC machine connection

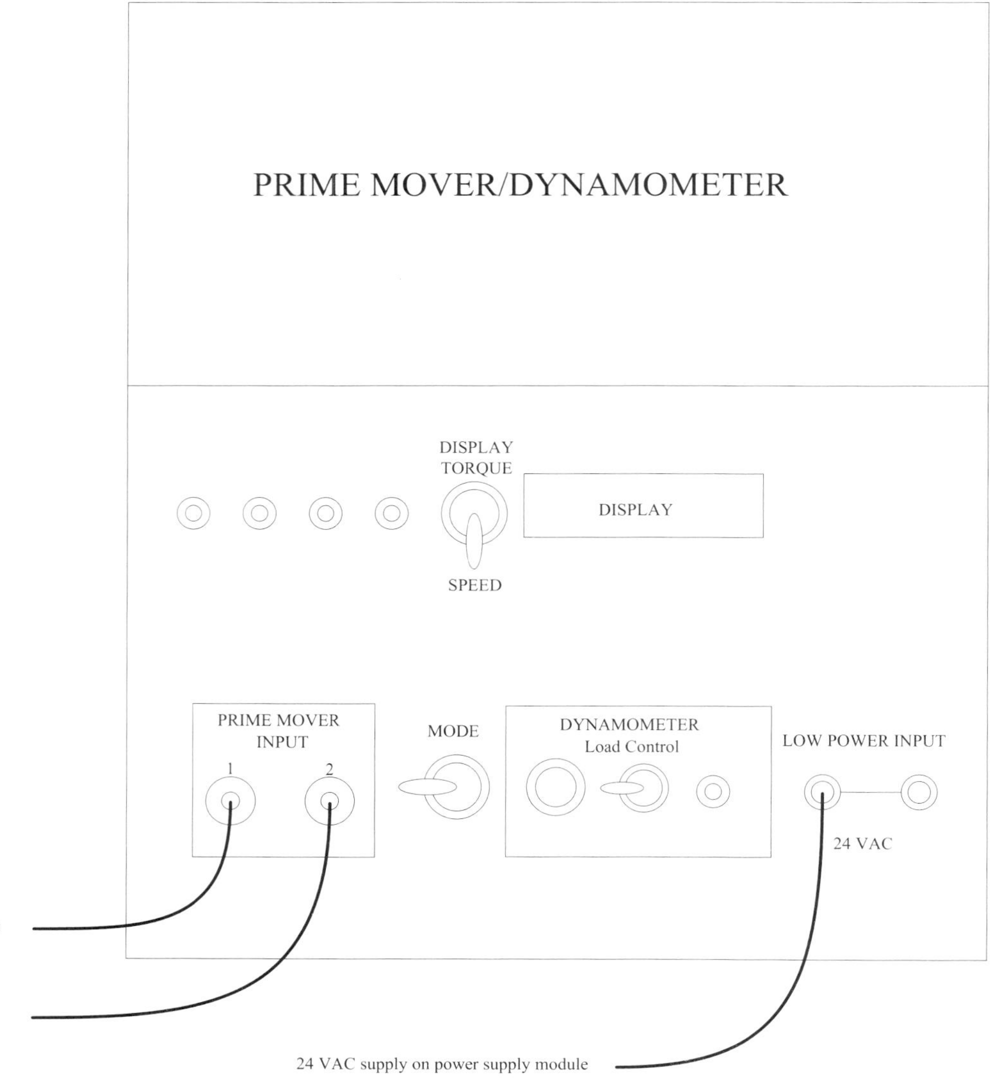

FIGURE 34-4 Connecting the prime mover/dynamometer module

7. Slowly increase the DC voltage applied to the armature with the voltage control knob located on the front of the power supply. Observe the direction of rotation of the DC machine. It should turn in a clockwise direction as you face the machine.
8. **Return the voltage to 0 V and turn off the power supply.** If the DC machine did *not* turn in the clockwise direction, change the armature leads, labeled 1 and 2, on the DC machine module.
9. Connect the circuit shown in Figure 34-5. Do not disconnect the DC machine or prime mover/dynamometer.
10. Check the field excitation switch located on the synchronous machine module. It should be in the open, or off, position.
11. Open all switches on the variable capacitance module.
12. Turn on the power supply and increase the voltage applied to the armature to 120 VDC. This value of voltage should be maintained throughout the experiment.

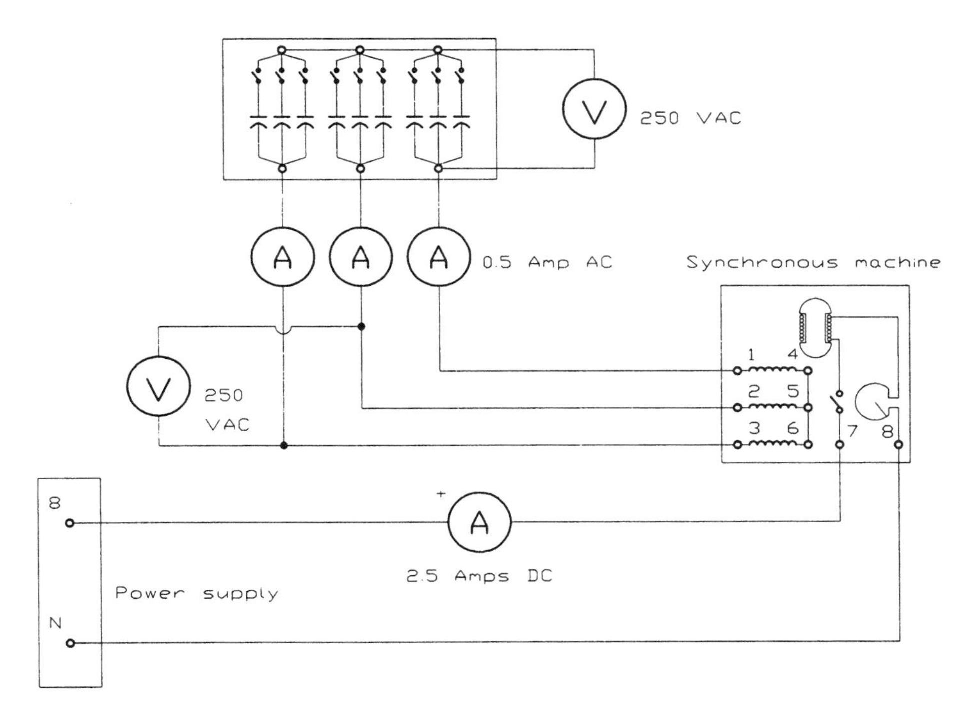

FIGURE 34-5 Three-phase alternator connection

13. Using the tachometer, set the speed of the DC machine to 1400 RPM by adjusting the shunt field rheostat located on the front of the DC machine module. If the prime mover/dynamometer module is being used, increase the voltage supplied to the input by turning the voltage control knob on the power supply until a speed of 1400 RPM is indicated.

14. Set the field control rheostat located on the front panel of the synchronous machine to the full counterclockwise position.

15. Excite the rotor of the alternator by closing the field excitation switch located on the front panel of the synchronous machine module.

16. Using the variable-capacitance load module, connect 4.4 μF of capacitance to each phase. (Close all three switches connected to the three 4.4-μF capacitors.)

17. Adjust the field control rheostat on the synchronous machine module until a voltage of 100 V is applied across each phase. Also recheck the speed of the DC machine. A speed of 1400 RPM should be maintained.

18. Measure the amount of current flow through the capacitors.

    I_C = ______________ A

19. The purpose of this part of the experiment is to show that the speed of rotation determines output frequency. Because the value of capacitance for each phase is known, 4.4 μF, the capacitive reactance will be proportional to the frequency. Before the frequency can be computed, however, the amount of capacitive reactance must be computed. Compute the amount of capacitive reactance using the formula

$$X_C = \frac{E_P}{I_C}$$

X_C = ______________ Ω

20. Compute the output frequency of the alternator using the formula

$$F = \frac{1}{2\pi CX_C}$$

NOTE: The value of capacitance is given in microfarads. This value must be converted to farads for use in the formula. To convert microfarads to farads, move the decimal point six places to the left.

$F =$ ______________ Hz

21. Reset the speed of the direct current machine or prime mover/dynamometer to 1600 RPM.
22. Adjust the phase voltage across each capacitor to 110 V.
23. Measure the amount of current flow through each phase.

$I_C =$ ______________ A

24. Compute the amount of capacitive reactance in the circuit.

$X_C =$ ______________ Ω

25. Compute the output frequency.

$F =$ ______________ Hz

26. Reset the speed of the DC machine or prime mover/dynamometer for a value of 1800 RPM.
27. Readjust the output voltage of the alternator for a value of 120 V per phase.
28. Measure the current flow through each phase.

$I_C =$ ______________ A

29. Compute the capacitive reactance.

$X_C =$ ______________ Ω

30. Compute the output frequency of the alternator.

$F =$ ______________ Hz

31. Open all switches on the variable-capacitance module.
32. Readjust the field control rheostat located on the front panel of the synchronous machine module to the full counterclockwise position.
33. Measure the amount of DC excitation current flowing through the rotor.

______________ A

34. Measure the line-to-line voltage of the alternator.

    ______________ V

35. Readjust the field control rheostat to the full clockwise position.
36. Measure the amount of DC excitation current flowing through the rotor.

    ______________ A

37. Measure the line-to-line voltage of the alternator.

    ______________ V

38. Notice that the amount of excitation current to the rotor controls the amount of output voltage of the alternator.
39. Open the field excitation switch on the synchronous machine module.
40. **Return the DC voltage applied to the armature to 0 V and turn off the power supply.**
41. Disconnect the circuit and return the components to their proper place.

Review Questions

1. What two factors determine the output frequency of an alternator?

   A. ____

   B. ____

2. At what speed must a six-pole alternator turn to produce 60 Hz?

   ______________ RPM

3. What three factors determine the output voltage of an alternator?

   A. ____

   B. ____

   C. ____

4. What are sliprings used for on an alternator?

   ____

   ____

5. Is the rotor excitation current AC or DC?

   ____

Exercise 35

Paralleling Alternators

Objectives

After completing this lab you should be able to:

- Discuss the reasons for paralleling alternators.
- Discuss the requirements for paralleling alternators.
- Connect and parallel two alternators.

Materials and Equipment

Power supply module	EMS 8821
AC ammeter module	EMS 8425
Direct current machine module	EMS 8211
or prime mover/dynamometer module	EMS 8960
Synchronous machine module	EMS 8241
DC metering module	EMS 8412
Synchronizing module	EMS 8621
AC voltmeter module	EMS 8426
Hand-held tachometer	EMS 8920
Variable-resistance module	EMS 8311

Discussion

Alternators are used to generate most of the electric power in the world. Because one alternator cannot produce all the power that is required, it often is necessary to use more than one machine. When more than one alternator is to be used, they are connected in parallel with each other. Three conditions must be met before alternators can be placed in parallel:

1. The phases must be connected in such a manner that the phase rotations of all the machines are the same.
2. Phases A, B, and C of one machine must be in sequence with phases A, B, and C of the other machine. For example, phase A of alternator 1 must reach its positive peak value of voltage at the same time that phase A of alternator 2 does, as shown in Figure 35-1.
3. The output voltage of the two alternators should be the same.

The most common method of detecting when the phase rotation of one alternator is matched to the phase rotation of the other alternator is with the use of three lights, as in Figure 35-2. In Figure 35-2, the two alternators that are to be paralleled are connected

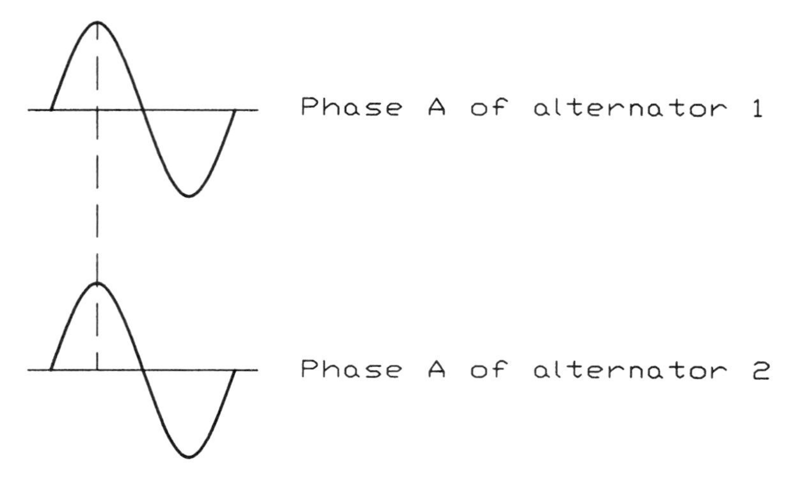

FIGURE 35-1 The two alternators must be in phase.

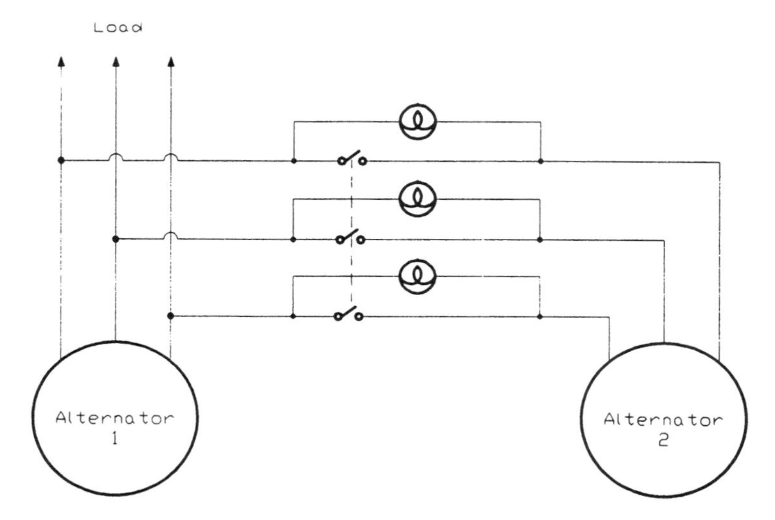

FIGURE 35-2 Lamps can be used to test for phase rotation.

together through a synchronizing switch. A set of lamps acts as a resistive load between the two machines when the switch contacts are in the open position. The voltage developed across the lamps is proportional to the difference in voltage between the two alternators. The lamps are used to indicate two conditions:

1. The lamps indicate when the phase rotation of one machine is matched to the phase rotation of the other machine. When both alternators are operating, both are producing a voltage. The lamps will blink on and off when the phase rotation of one machine is not synchronized to the phase rotation of the other machine. If all three lamps blink on and off at the same time, the phase rotation of alternator 1 is correctly matched to the phase rotation of alternator 2. If the lamps blink on and off in an alternate sequence, the phase rotation between the two machines is not correctly matched, and two lines of alternator 2 should be switched.
2. The lamps also indicate when the phase of one machine is synchronized with the phase of the other machine. If the positive peak of alternator 1 does not occur at the same time as the positive peak of alternator 2, there will be a potential between the two machines. This will permit the lamp to glow. The brightness of the lamp indicates how far out of synchronism the two machines are. When the peak voltages of the two alternators occur at the same time, there is no potential difference between them. The lamps should be off at this time. The synchronizing switch should never be closed when the lamps are glowing.

Another instrument often used for paralleling two alternators is the synchroscope. The synchroscope compares the phase angle difference between two phases of each alternator. When the two phases are synchronized, the paralleling switch is closed. If a synchroscope is not available, the two alternators can be paralleled using three lamps as described earlier. If the three-lamp method is used, an AC voltmeter connected across the same phase of each machine will indicate when the potential difference between the two machines is zero. That is the point at which the paralleling switch should be closed.

After the alternators have been paralleled together, the power input to alternator 2 must be increased to permit it to share part of the load. For example, if the alternator were being driven by a steam turbine, the power of the turbine would have to be increased. The device used to supply the turning force to the alternator is known as the prime mover.

Name ______________________________ Date ______________

Procedure

1. Connect the circuit shown in Figure 35-3 if the DC machine is being used. If the prime mover is being used, connect it as shown in Figure 35-4.
2. If the DC machine is being used as the prime mover, turn on the power supply and adjust the shunt field rheostat, located on the front of the DC machine module, until a current of 0.4 A flows through the shunt field. If the prime mover/dynamometer is being used, turn on the power supply and proceed to Step 3.
3. Slowly increase the DC voltage applied to the armature of the DC machine or prime mover/dynamometer with the voltage control knob located on the power supply. The motor should turn in a clockwise direction.
4. **Return the voltage to 0 V and turn off the power supply**. If the motor did not turn in the clockwise direction, reverse the armature leads, 1 and 2, on the direct current machine module.
5. Connect the circuit shown in Figure 35-5. Do not disconnect the DC machine or prime mover/dynamometer.
6. Check the synchronizing module shown in Figure 35-6 to make sure that the paralleling switch is open.
7. Set the variable-resistance module to have a resistance of 300 Ω per phase. (Each of the three 300-Ω switches should be closed, or turned on.)
8. If the DC machine is being used as the prime mover, turn on the power supply and adjust the DC voltage for 120 volts. This voltage value should be maintained throughout the experiment. If the prime mover/dynamometer is being used, turn on the power supply and proceed to Step 9.
9. If the DC machine is being used, adjust the speed for 1700 RPM by using the tachometer and shunt field rheostat. If the prime mover/dynamometer is being used, set the prime mover/dynamometer for a speed of 1700 RPM by turning the voltage control knob located on the power supply.

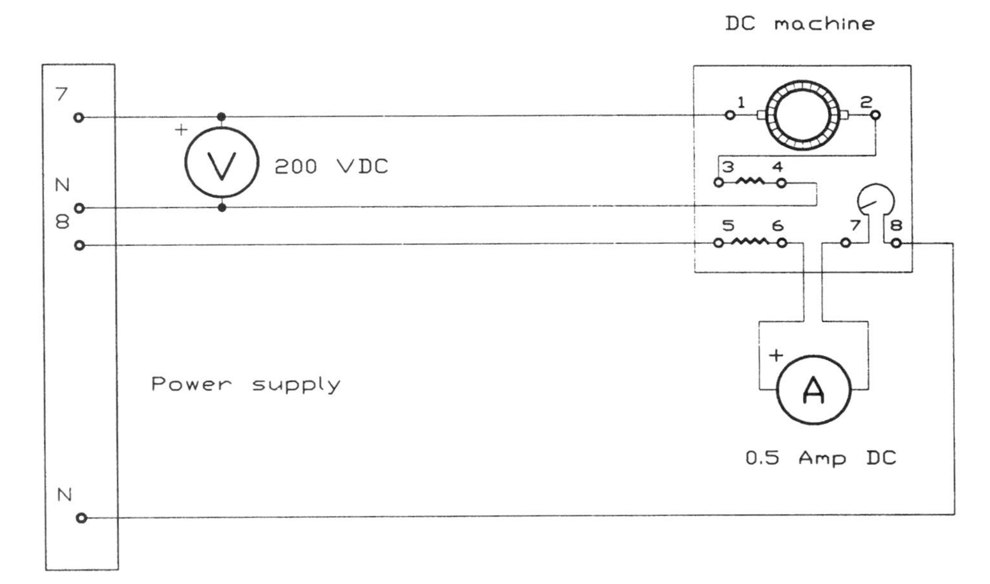

FIGURE 35-3 DC machine connection

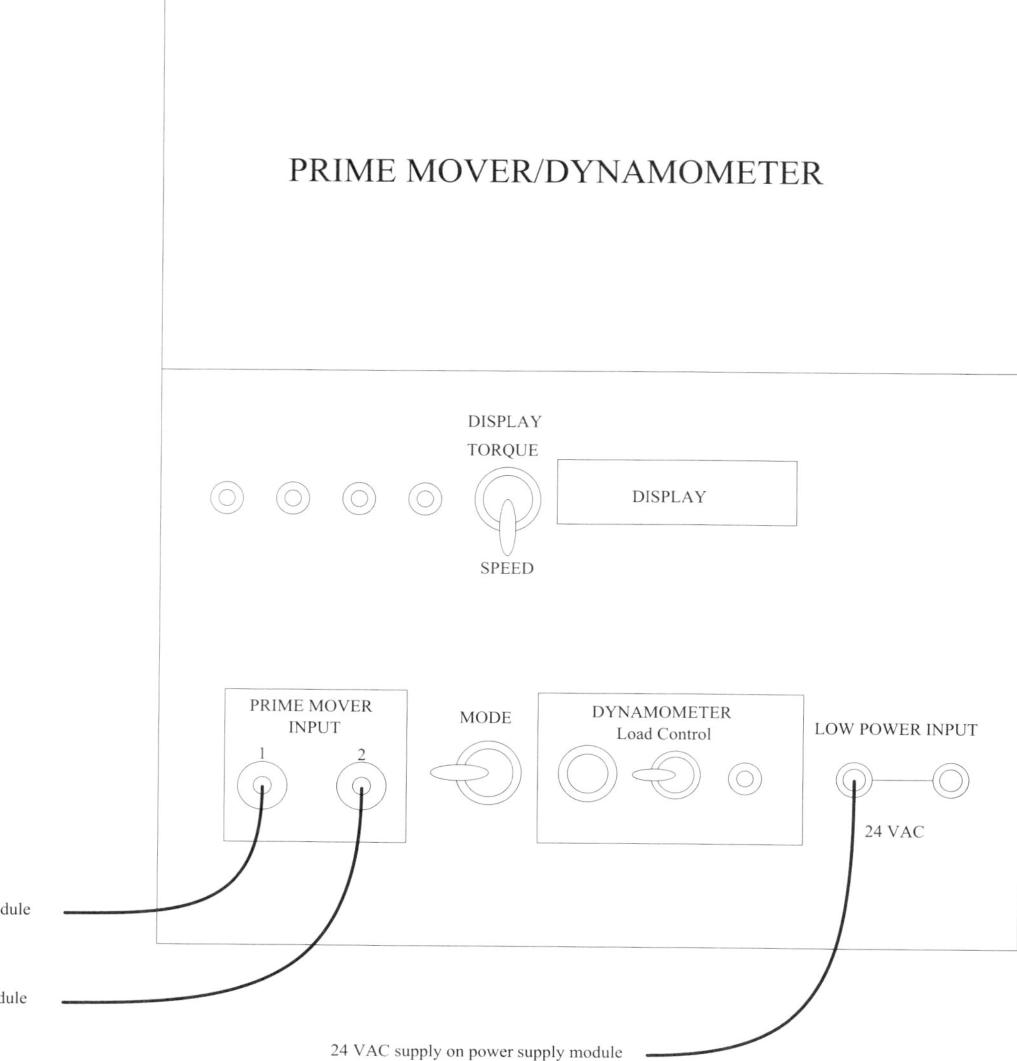

FIGURE 35-4 Connecting the prime mover/dynamometer module

10. Measure the AC current being applied to the resistance load.

    I_R = ______________ A

11. Excite the rotor of the alternator by closing the excitation switch located on the front of the synchronous machine module.

12. Adjust the rheostat on the synchronous machine module until the alternator has an output line voltage of 208 V.

13. The lights on the front of the synchronizing module should be blinking on and off at this point. These lights are used to show the phase rotation relationship between the incoming power and the alternator. If the phase rotation is correct, all the lamps will be blinking on and off at the same time. If the rotation is incorrect, the lamps will be blinking on and off in alternate sequence. Are the lamps blinking on and off at the same time or in alternate sequence?

    ____

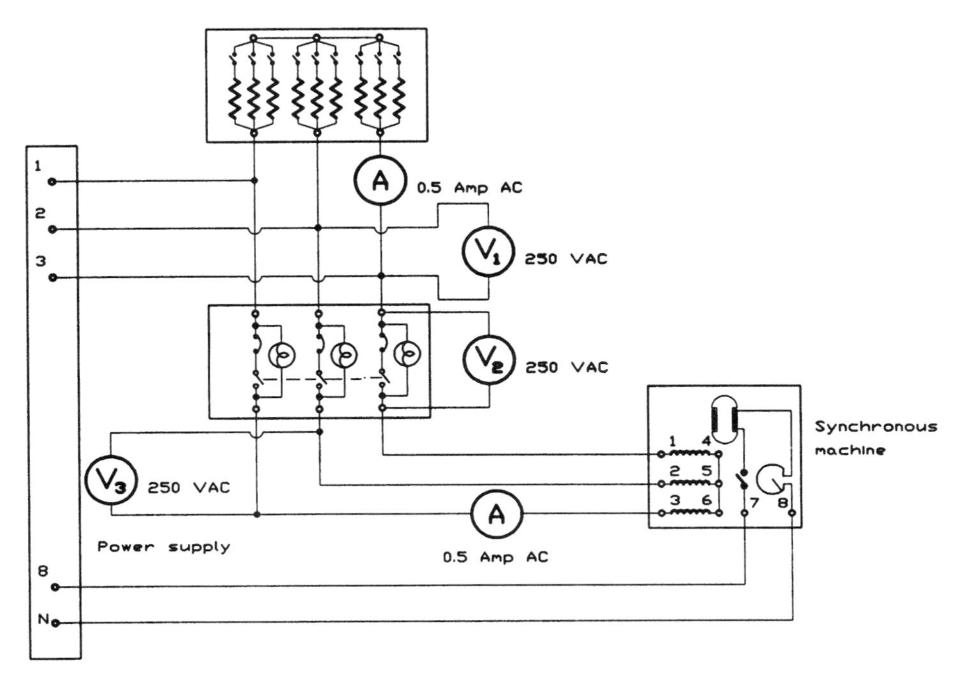

FIGURE 35-5 Connection of the synchronous machine to the power source

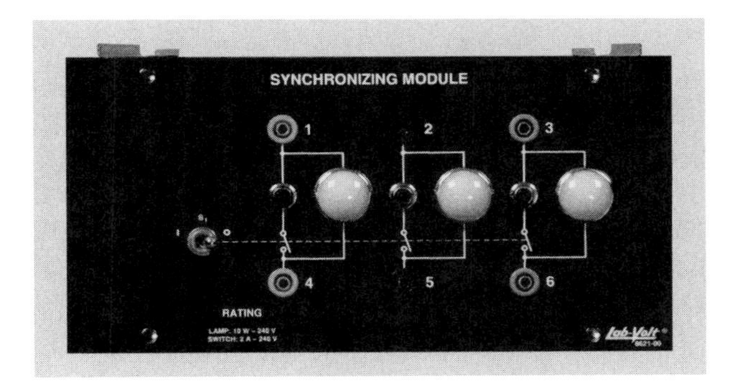

FIGURE 35-6 Synchronizing module (Courtesy of Lab-Volt® Systems, Inc.)

14. **Turn off the voltage output of the alternator by opening the excitation switch located on the front panel of the synchronous machine.** Change any two of the output leads of the alternator.

15. Close the excitation switch located on the front panel of the synchronous machine.

16. Are the three lamps located on the synchronizing module blinking on and off at the same time or in alternate sequence?

____

17. Set the output leads of the alternator to the position required to cause the lamps on the synchronizing module to blink on and off at the same time. If it is necessary to change the alternator leads, open the excitation switch located on the front of the synchronous machine module first. Reclose the switch after the leads have been changed.

18. Adjust the speed of the direct current machine or the prime mover/dynamometer until the lamps are blinking on and off at a very slow rate. Observe the voltmeter connected across the phase of the incoming power and the phase of the alternator (V_2). This meter is used to indicate the voltage difference between the two power sources.
19. When the voltage between the two power sources drops to its lowest point, close the paralleling switch located on the front of the synchronizing module. This will parallel the alternator to the incoming power line.
20. Measure the current flow at the output of the alternator.

    I = ______________ A
21. Slowly increase the speed of the direct current machine or prime mover/ dynamometer and observe the ammeter connected to the alternator. It may be necessary to increase the voltage supplied to the armature of the DC machine or prime mover/dynamometer. Do not exceed 130 V.
22. Measure the current flow at the output of the alternator.

    I = ______________ A
23. The increase of alternator current indicates that power is being supplied to the incoming power lines by the alternator.
24. **Open the paralleling switch and turn off the power supply.**
25. Disconnect the circuit and return the components to their proper place.

Review Questions

1. What conditions must be met before two alternators can be paralleled together?

   A. ____

   B. ____

   C. ____
2. How can the phase rotation of one alternator be changed in relationship to the other alternator?

   ____
3. What are the functions of the synchronizing lamps?

   A. ____

   B. ____
4. What is a synchroscope?

   ____
5. Assume that alternator A is supplying power to a load and that alternator B is to be paralleled to A. After the paralleling has been completed, what must be done to permit alternator B to share the load with alternator A?

   ____

Exercise 36

The Squirrel Cage Induction Motor

Objectives

After completing this lab you should be able to:

- Discuss the operation of a three-phase squirrel cage induction motor.
- Calculate horsepower and efficiency of an induction motor.
- Connect a three-phase induction motor and make measurements with test instruments.

Materials and Equipment

Squirrel cage induction motor module	EMS 8221
Power supply module	EMS 8821
Electrodynamometer module	EMS 8911
or prime mover/dynamometer	EMS 8960
AC ammeter module	EMS 8425
Hand-held tachometer	EMS 8920
Three-phase wattmeter module	EMS 8441

Discussion

The squirrel cage motor receives its name from the type of rotor used in the motor. A squirrel cage rotor is made by connecting bars to two end rings. The result looks very similar to a squirrel cage. A squirrel cage is a cylindrical device constructed of heavy wire. A shaft is placed through the center of the cage. This permits the cage to spin around the shaft. A squirrel cage is placed inside the cage of small pets such as squirrels and hamsters to permit them to exercise by running inside the cage.

The term ***induction motor*** indicates that the motor operates by means of induction. When power is connected to the stator, a rotating magnetic field is established inside the motor. This rotating field cuts through the bars of the squirrel cage rotor and induces a voltage into them. Because the bars are shorted together at each end, a current flows through the bars. When current flows through the bars, it produces a magnetic field around the rotor. The magnetic field of the rotor is attracted to the rotating magnetic field of the stator. This magnetic attraction causes the rotor to turn in the direction of the rotating field.
The speed of the rotating magnetic field is known as the synchronous speed of the motor. Synchronous speed is determined by two factors:

1. the number of stator poles and
2. the frequency of the supply voltage.

The chart below shows the synchronous speed for different numbers of stator poles at 60 Hz.

Number of stator poles	Speed (RPM)
2	3600
4	1800
6	1200
8	900

The synchronous speed can also be computed by using the formula

$$S = \frac{120\ F}{P}$$

where S = speed in RPM
F = frequency in Hz
P = number of poles per phase

The amount of torque produced by an induction motor is proportional to three factors:

1. the strength of the magnetic field of the stator,
2. the strength of the magnetic field of the rotor, and
3. the phase angle difference between the current in the stator and the current in the rotor.

NOTE: Maximum torque is developed when the flux in the rotor is in phase with the flux in the stator.

The direction of rotation of a three-phase motor can be reversed by changing any two of the stator windings. This reverses the direction of the rotating magnetic field.

MOTOR CHARACTERISTICS

When a squirrel cage motor is first started, it will have a current draw several times greater than its normal running current. The actual amount of starting current is determined by the type of rotor bars, the horsepower rating of the motor, and the applied voltage. The type of rotor bars is indicated by the code letter found on the nameplate of a squirrel cage motor. Table 430.7(b) of the National Electrical Code can be used to compute the locked rotor current (starting current) of a squirrel cage motor when the horsepower and code letter are known. This large starting current is caused by the fact that the rotor is not turning when power is first applied to the stator. Because the rotor is not turning, the squirrel cage bars are cut by the rotating magnetic field at a fast rate. Recall that one of the factors that determines the amount of induced voltage is the speed of the cutting action. This high induced voltage causes a large amount of current to flow in the rotor. The large current flow in the rotor causes a large amount of current flow in the stator. Because the stator induces a current into the rotor, the stator acts as the primary of a transformer. The rotor functions as the secondary of a transformer. Because large amounts of current flow in the stator and rotor, a strong magnetic field is established in both the stator and rotor. The starting torque of a squirrel cage motor is high because the magnetic field of both the stator and rotor are strong at this point. Recall, however, that the third factor for determining the torque developed by an induction motor is the difference in phase angle between stator current and rotor current. The rotor is being cut at a high rate of speed by the rotating stator field, so the bars in the squirrel cage rotor appear to be very inductive at this point because of the high frequency of the induced voltage. This causes the phase angle difference between rotor and stator current to be great, causing the rotor and stator flux to be out of phase with each other. Although the starting torque of a squirrel cage motor is high, it does not develop as much starting torque per amp of starting current as does the wound rotor induction motor.

As the rotor increases in speed, the squirrel cage bars are cut by the rotating field at a slower rate. This causes less current flow in the rotor and, therefore, less current flow in the stator.

If the motor is operating at no load, the rotor will accelerate to a point close to synchronous speed. The rotor of an induction motor can never reach synchronous speed because at that point there would be no induced voltage in the rotor. If there were no induced voltage in the rotor, there could be no current flow in the rotor and consequently no rotor magnetic field. Without a rotor magnetic field, no torque could be developed to operate the motor.

Now assume that a load is added to the motor. This load causes the rotor to slow down. When the rotor slows, the rotating magnetic field cuts the rotor bars at a faster rate, which induces more current into the rotor. This added current causes a stronger magnetic field to be established, which results in an increase in torque. It should also be noted that because the rotor is turning at close to synchronous speed, voltage is being induced in the rotor at a low frequency. This causes the rotor circuit to be more resistive than inductive. The rotor bars now appear to be more resistive, and the current in the rotor is almost in phase with the stator current. This results in the rotor and stator flux being almost in phase with each other, producing a strong running torque.

MOTOR CALCULATIONS

In the following example, output horsepower and motor efficiency will be computed. It is assumed that a 1/2-horsepower squirrel cage motor is connected to a load. A wattmeter is connected to the motor, and the load is calibrated in pound-inches of torque. The motor is operating at a speed of 1725 RPM and producing a torque of 16 lb-in. The wattmeter is indicating an input power of 500 W.

The actual amount of horsepower being produced by the motor can be calculated by using the formula

$$hp = \frac{6.28 \times RPM \times L \times P}{33,000}$$

where hp = horsepower
6.28 is a constant
RPM = speed in revolutions per minute
L = distance in feet
P = pounds of force
33,000 is a constant (550 lb ft. × 60 s)

Because the formula uses feet for the distance and the torque of the motor is rated in pound-inches, L will be changed to 1/12 of a foot, or 1 inch. To simplify the calculation, the fraction 1/12 will be changed into its decimal equivalent (0.08333). To compute the output horsepower, substitute the known values in the formula.

$$hp = \frac{6.28 \times 1725 \times 0.08333 \times 16}{33,000}$$

$$hp = \frac{14,444}{33,000}$$

$$hp = 0.438 \text{ hp}$$

One horsepower is equal to 746 W. The output power of the motor can be computed by multiplying the output horsepower value by 746.

$$\text{power out} = 746 \times 0.438 \text{ hp}$$

$$\text{power out} = 326.5 \text{ W}$$

The efficiency of the motor can be computed by using the formula

$$\text{eff.} = \frac{\text{power out}}{\text{power in}} \times 100$$

$$\text{eff.} = \frac{326.5 \text{ W}}{500 \text{ W}} \times 100$$

$$\text{eff.} = 0.653 \text{ W} \times 100$$

$$\text{eff.} = 65.3\%$$

Name ______________________________ Date ______________

Procedure

1. Connect the circuit shown in Figure 36-1 if the electrodynamometer is being used. The squirrel cage induction motor module is shown in Figure 36-2. If the prime mover/dynamometer module is being used instead of the electrodynamometer, connect the 24 VAC low-voltage supply to the power supply. **Do not** connect the squirrel cage motor and dynamometer together with the timing belt at this time.
2. Fill in the chart shown in Figure 36-3 for each load shown. The measured values of load torque and RPM are used to compute horsepower. The horsepower can be used to compute output power. The values of output power and input power are used to compute efficiency for the motor. The first measurement is to be made with the motor and electrodynamometer disconnected.
3. Turn on the power supply and check the direction of rotation of the motor.
4. **Turn off the power supply and reverse any two of the stator leads.**
5. Turn on the power supply and again check the direction of rotation of the motor.
6. Did the direction of rotation reverse when the stator leads were changed?

   ____

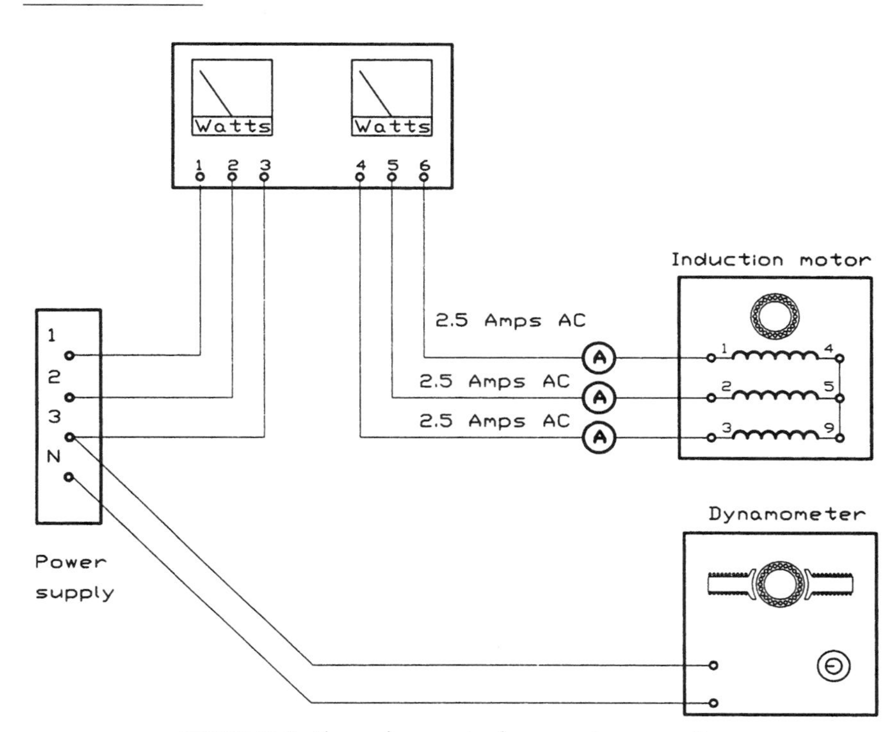

FIGURE 36-1 Three-phase squirrel cage motor connection

FIGURE 36-2 Squirrel cage induction motor (Courtesy of Lab-Volt® Systems, Inc.)

Load torque lb-in.	Armature current (amps)	Speed (RPM)	Input power (watts)	Output horsepower (hp)	Output power (watts)	Eff. (%)
0						
2						
4						
6						
8						

FIGURE 36-3

7. If the motor is not turning in a clockwise direction as you face the motor, turn off the power supply and change two stator leads. If the motor is operating in a clockwise direction, fill in the first set of values for no-load in the chart.
8. **Turn off the power supply** and connect the timing belt between the squirrel cage induction motor and the dynamometer.
9. Turn on the power supply and fill in the remaining values in the chart by adjusting the electrodynamometer for each of the loads listed.
10. **Turn off the power supply and disconnect the circuit**. Return the components to their proper place.

Review Questions

1. Did the efficiency of the motor increase or decrease as load was added?

   ____

2. Did the motor current increase or decrease as load was added?

   ____

3. Did the motor RPM increase or decrease as load was added?

   ____

4. Why does a decrease of motor RPM cause the motor current to increase?

   ____

5. What is synchronous speed?

   ____

6. What two factors determine synchronous speed?

   A. ____

   B. ____

7. What three factors determine the amount of torque developed by an induction motor?

   A. ____

   B. ____

   C. ____

8. Why can an induction motor never operate at synchronous speed?

   ____

9. A squirrel cage induction motor is operating at 1175 RPM and producing a torque of 22 lb-ft. What is the horsepower output of the motor?

   ______________ hp

10. A wattmeter measures the input power of the motor in problem 9 to be 565 W. What is the efficiency of the motor?

    ______________ %

Exercise 37

Power Factor Correction For Three-Phase Motors

Objectives

After completing this lab you should be able to:

- Discuss power factor.
- Determine the power factor of a motor.
- Calculate the amount of capacitance needed for correcting motor power factor.

Materials and Equipment

Squirrel cage induction motor module	EMS 8221
Power supply module	EMS 8821
Variable capacitance module	EMS 8331
AC ammeter module	EMS 8425
Three-phase wattmeter module	EMS 8441
Electrodynamometer module	EMS 8911
or prime mover/dynamometer	EMS 8960

Discussion

Circuit power factor can be very important in an industrial environment. Power factor is the ratio between the apparent power, volt-amperes, and the true power, watts. The power company must produce, or generate, enough power to supply the apparent power. For this reason, utility companies often charge an extra high rate when the power factor of a large business becomes too low.

In Figure 37-1, an induction motor is connected to the power line. Two wattmeters have been connected to the motor to provide a true power measurement. Ammeters have also been connected in series with the motor to provide a measurement of current. The motor is connected to a 480-V line. When the motor is in operation, the ammeters indicate a current flow of 27 A, and the wattmeter connection indicates a true power in the circuit of 14,140 W.

The apparent power can be computed using the formula

$$VA = \sqrt{3} \times E_L \times I_L$$

$$VA = 1.732 \times 480\ V \times 27\ A$$

$$VA = 22,446.7$$

The power factor for this circuit can be calculated using the formula

$$PF = \frac{W}{VA} \times 100$$

$$PF = \frac{14140\ W}{22446.7\ VA} \times 100$$

$$PF = 0.63 \times 100$$

$$PF = 63\%$$

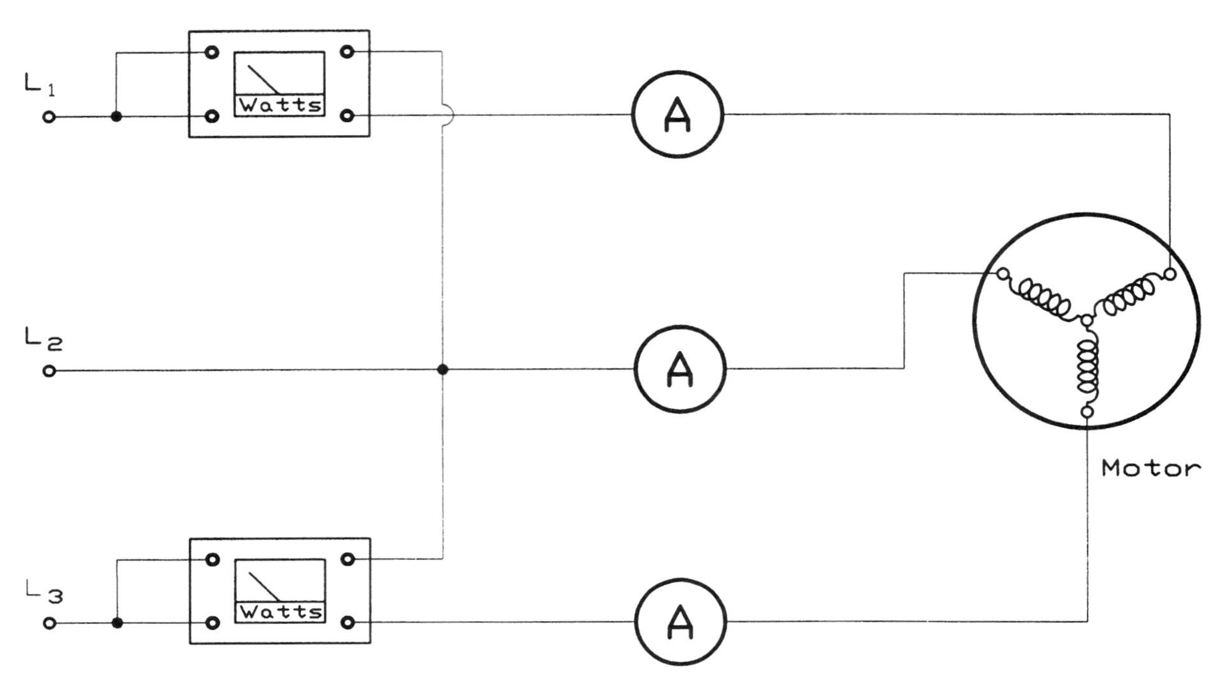

FIGURE 37-1 Using single-phase wattmeters to measure three-phase power

To correct the power factor of this circuit, you must compute the amount of reactive power (VARs). There is only one path for current flow through the phase windings of the motor, so the resistive and inductive parts of the circuit appear to be connected in series with each other. The formula used to compute VARs will be

$$\text{VARs} = \sqrt{\text{VA}^2 - \text{W}^2}$$
$$\text{VARs} = 17,433$$

Because this circuit is inductive, the VARs is inductive. For the power factor to be corrected to 100%, an equal amount of capacitive VARs must be connected in parallel with the motor. It is generally impractical to correct the power factor to 100%, however, so in this example the power factor will be corrected to 95%. In order to do this, we will have to compute the inductive VARs needed to produce a power factor of 95%.

The first step in solving this problem is to determine the amount of apparent power needed to produce a power factor of 95%. This can be done by dividing the true power by the desired power factor.

$$\text{VA} = \frac{\text{W}}{0.95}$$
$$\text{VA} = \frac{14140\ \text{W}}{0.95}$$
$$\text{VA} = 14884.2$$

Now that the desired apparent power is known, the inductive VARs needed to produce this value of volt-amps can be computed.

$$\text{VARs} = \sqrt{\text{VA}^2 - \text{W}^2}$$

$$\text{VARs} = \sqrt{14884.2^2 - 14140^2}$$

$$\text{VARs} = 4647.5$$

The capacitive VARs needed to reduce the inductive VARs to this value can be computed by subtracting the desired value of inductive VARs from the existing value.

$$\text{capacitive VARs needed} = 17,433 - 4647.5$$

$$\text{capacitive VARs} = 12,785.4$$

To compute the amount of capacitance needed, we must first compute the amount of current that must flow in the capacitor circuit. This can be done by using the formula

$$I_C = \frac{\text{VARs}}{E_L \times 1.732}$$

$$I_C = \frac{12,785.4 \text{ VARs}}{480 \text{ V} \times 1.732}$$

$$I_C = 15.38 \text{ A}$$

The amount of capacitive reactance needed to produce this current flow can now be computed using the formula

$$X_C = \frac{E_{PHASE}}{I_C}$$

$$X_C = \frac{277 \text{ V}}{15.38 \text{ A}}$$

$$X_C = 18\ \Omega$$

Notice that the phase value of voltage is used instead of the line value. This was done because the capacitors are to be connected in a wye connection and the amount of voltage drop across each capacitor must be used when determining the amount of capacitive reactance.

The amount of capacitance needed to produce this amount of capacitive reactance can be calculated using the formula

$$C = \frac{1}{2\pi FX_C}$$

$$C = 0.000147 \text{ F} = 147\ \mu\text{F}$$

Three capacitors, each having a value of 147 μF, will be connected in a wye connection. This wye connection will then be connected in parallel with the motor.

Name ______________________________ Date ______________

Procedure

1. Connect the circuit shown in Figure 37-2 if the electrodynamometer is being used. If the prime mover/dynamometer module is being used instead of the electrodynamometer, connect the 24 VAC low-voltage supply to the power supply.
2. Connect the motor and dynamometer together with the timing belt.
3. Turn on the power supply and check the direction of rotation of the motor. The motor should turn in the clockwise direction as you face the motor. If it does not, turn off the power supply and change any two of the motor stator leads. Then turn the power supply on again.
4. Adjust the dynamometer until there is a load of 8 lb-in. on the motor.
5. Using the three-phase wattmeter, measure and record the true power in the circuit.

   P = ______________ W

6. Measure and record the amount of current flow to the motor.

   I = ______________ A

7. Compute the amount of apparent power using the formula

$$VA = \sqrt{3} \times E_{LINE} \times I_{LINE}$$

   VA = ______________

8. Compute the power factor of this circuit using the formula

$$PF = \frac{W}{VA} \times 100$$

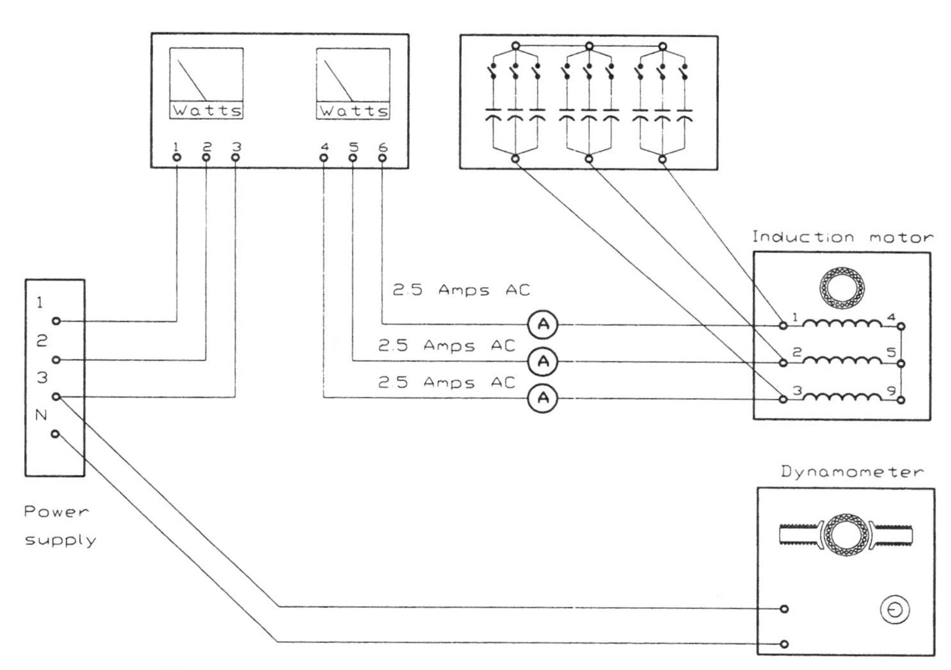

FIGURE 37-2 Correcting the power factor for a squirrel cage motor

9. Calculate the amount of reactive power in this circuit using the formula

$$\text{VARs} = \sqrt{\text{VA}^2 - \text{W}^2}$$

10. Compute the apparent power needed to correct the circuit power factor to 95% by dividing the true power by the desired power factor.

$$\text{VA} = \frac{\text{W}}{0.95}$$

Desired VA = ______________

11. Compute the amount of reactive power need to produce the desired value of apparent power using the formula

$$\text{VARs}_{(\text{DESIRED})} = \sqrt{\text{VA}_{\text{DESIRED}}{}^2 - \text{W}^2}$$

$\text{VARs}_{(\text{DESIRED})}$ = ______________

12. To find the capacitive VARs needed, subtract the desired VARs from the existing VARs.

$$\text{Capacitive VARs} = \text{VARs}_{(\text{EXISTING})} - \text{VARs}_{(\text{DESIRED})}$$

Capacitive VARs = ______________

13. Compute the amount of capacitive current needed using the formula

$$I_C = \frac{\text{VARs}}{E_{\text{LINE}} \times 1.732}$$

I_C = ______________ A

14. Compute the amount of capacitive reactance needed using the formula

$$X_C = \frac{E_{\text{PHASE}}}{I_C}$$

X_C = ______________ Ω

15. Calculate the amount of capacitance needed using the formula

$$C = \frac{1}{2\pi f X_C}$$

C = ______________ μF

16. Using the capacitive load module, connect this amount of capacitance in the circuit. It may not be possible to connect the exact amount of capacitance because of the fixed values of the load module. An effort should be made to make the capacitance value as close as possible. It is generally preferred to have the power factor slightly lagging rather than leading. For this reason, if the exact amount of capacitance cannot be connected, use an amount of capacitance that is less than the computed value instead of greater than the computed value.

17. Measure and record the true power of the circuit.

    P = ______________ W

18. Measure and record the amount of current flow to the motor.

    I = ______________ A

19. Compute the apparent power of the circuit using the formula

$$VA = \sqrt{3} \times E_{LINE} \times I_{LINE}$$

    VA = ______________

20. Compute the power factor of this circuit using the formula

$$PF = \frac{W}{VA} \times 100$$

    PF = ______________ %

21. Compare this power factor value with the previous value.
22. **Turn off the power supply and disconnect the circuit.**
23. Return the components to their proper location.

Review Questions

1. What is power factor?

    ____

2. A motor is connected to a 240-V, three-phase line. The true power of the motor is 2793 W. The motor has a current draw of 12 A. Compute the following values for this circuit.

    VA = ______________

    PF = ______________ %

    VARs = ______________

3. Using the information in the above problem, find the amount of capacitance needed to correct the power factor to a value of 95%.

    C = ______________ μF

Exercise 38

The Wound Rotor Induction Motor

Objectives

After completing this lab you should be able to:

- Discuss the operation of a wound rotor motor.
- Connect a wound rotor motor and make measurements of voltage, current, speed, and horsepower.

Materials and Equipment

Wound rotor motor module	EMS 8231
Power supply module	EMS 8821
AC ammeter module	EMS 8425
AC voltmeter module	EMS 8426
Three-phase wattmeter module	EMS 8441
Three-phase rheostat module	EMS 8731
Electrodynamometer module	EMS 8911
or prime mover/dynamometer	EMS 8960
Hand-held tachometer	EMS 8920

Discussion

The wound rotor induction motor is very popular in industry because of its high starting torque and low starting current. The stator winding of the wound rotor motor is the same as that of the squirrel cage motor. The difference between the two motors lies in the construction of the rotor. Recall that the squirrel cage rotor is constructed of bars connected together at each end by a shorting ring as shown in Figure 38-1.

The rotor of a wound rotor motor is constructed by winding three separate coils on the rotor 120° apart. These coils are then connected to three sliprings located on the rotor shaft as shown in Figure 38-2. Brushes connected to the sliprings provide external

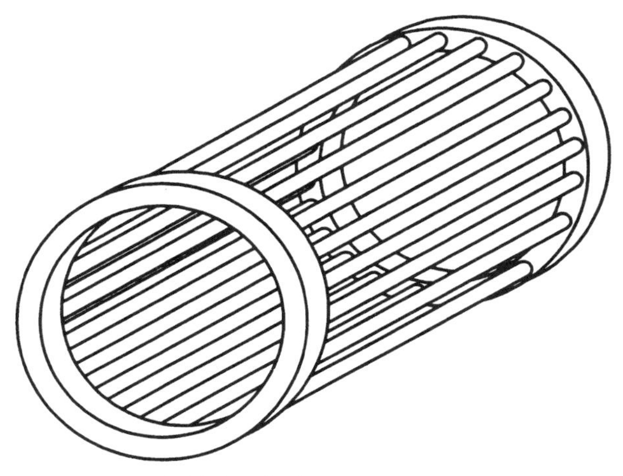

FIGURE 38-1 Squirrel cage rotor

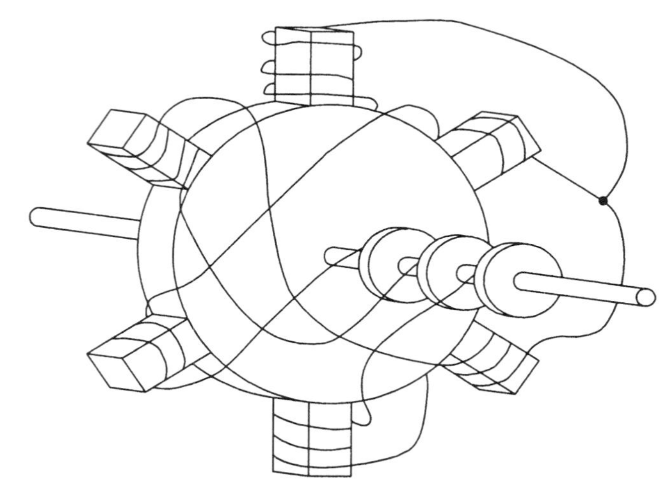

FIGURE 38-2 Wound rotor with sliprings

connection to the rotor circuit. This permits the rotor circuit to be connected to a set of resistors as shown in Figure 38-3.

The stator terminal connections are generally labeled T_1, T_2, and T_3. The rotor connections are commonly labeled M_1, M_2, and M_3. The schematic symbol for a wound rotor motor is shown in Figure 38-4.

MOTOR OPERATION

When power is applied to the stator winding, a rotating magnetic field is created in the motor. This magnetic field cuts through the windings of the rotor and induces a voltage into them. The amount of voltage induced in the rotor windings is determined by three factors:

1. the strength of the magnetic field in the stator,
2. the number of turns of wire in the rotor, and
3. the speed of the cutting action.

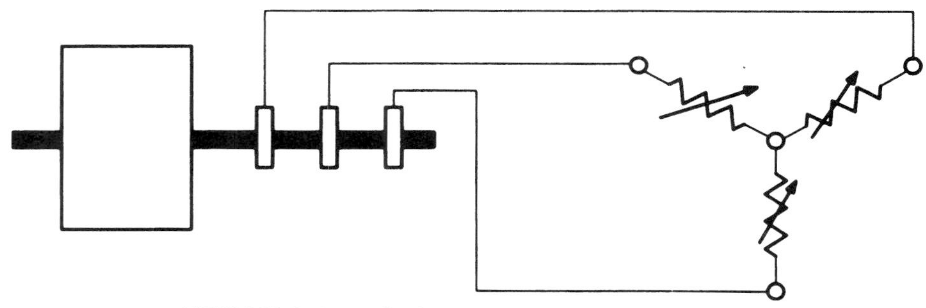

FIGURE 38-3 Rotor circuit connected to external resistance

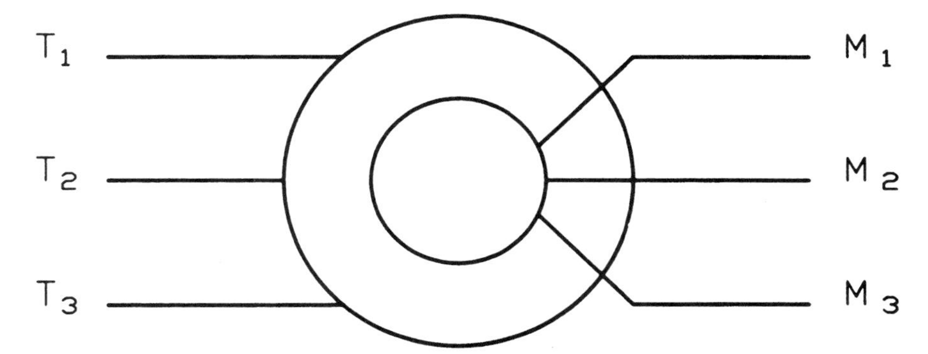

FIGURE 38-4 Schematic symbol used to represent a wound rotor induction motor

The amount of current flow in the rotor is determined by the amount of induced voltage and the total impedance of the rotor circuit (I = E/Z). When current flows through the rotor, a magnetic field is produced. This magnetic field is attracted to the rotating magnetic field of the stator.

As the rotor speed increases, the induced voltage decreases because of less cutting action between the rotor windings and rotating magnetic field. This produces less current flow in the rotor and, therefore, less torque. If rotor circuit resistance is reduced, more current can flow, which will increase motor torque, and the rotor will increase in speed. This action continues until all external resistance has been removed from the rotor circuit and the motor is operating at maximum speed. At this point, the wound rotor motor is operating in the same manner as a squirrel cage motor.

STARTING CHARACTERISTICS

The wound rotor motor will have a lower starting current per horsepower than a squirrel cage motor. The starting current is less because resistance is connected in the rotor circuit during starting. This resistance limits the amount of current that can flow in the rotor circuit. Since the stator current is proportional to rotor current because of transformer action, the stator current is less also.

Although the overall starting torque for a wound rotor motor is less than that of an equivalent horsepower squirrel cage motor, the starting torque is higher per amp of starting current than that of the squirrel cage motor. The starting torque is higher per amp of starting current because of the resistance in the rotor circuit. Recall that one of the factors that determines motor torque is the phase angle difference between the stator current and the rotor current. Since resistance is connected in the rotor circuit, the current flow in the rotor is closer to being in phase with the stator current than in a squirrel cage rotor. This produces a higher starting torque per amp of starting current for the wound rotor induction motor.

Name ______________________________ Date ______________

Procedure

1. Remove the wound rotor motor module Figure 38-5 from the mobile console.
2. Observe the stator winding around the inside of the motor housing.
3. Notice the windings of the rotor. Locate the sliprings at the rear of the motor. How many brushes are connected to the sliprings?

   ____

4. Trace the brush connections to the front panel of the module. Notice that the schematic drawing on the front panel indicates that the windings are connected together at one end. Are the rotor windings connected as a wye or as a delta?

   ____

5. Replace the wound rotor motor module in the console and connect the circuit shown in Figure 38-6 if using the electrodynamometer. If the prime mover/dynamometer is being used, connect the 24 VAC low-voltage connection to the power supply. Connect the motor and dynamometer together with the timing belt. Set the dynamometer control knob to the full counterclockwise position.

NOTE: In this experiment the open circuit voltage of the rotor will be measured. The Lab-Volt® EMS equipment is designed to be able to make this test without damage to the equipment. In the field, however, power should never be connected to the stator of a wound rotor motor when the rotor circuit is open. This can cause a high voltage to be induced into the rotor circuit and permanently damage the rotor.

6. Turn on the power supply and measure the amount of voltage being induced into the rotor.

   ____________ V

7. **Turn off the power supply.**

FIGURE 38-5 Wound rotor induction motor (Courtesy of Lab-Volt® Systems, Inc.)

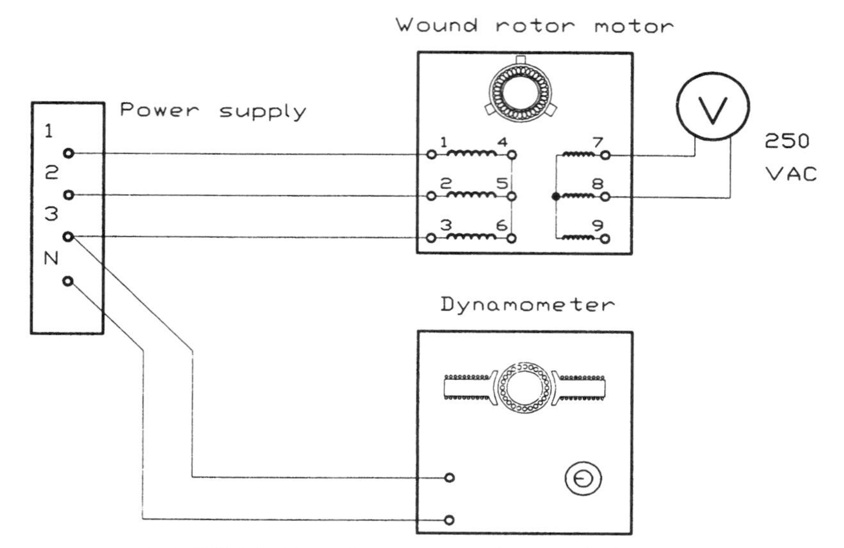

FIGURE 38-6 Measuring the induced voltage in the rotor

8. Notice that the rotor of the motor does not turn when power is connected. Since the rotor circuit is open, no current flows through the rotor circuit. Because there is no current flow through the rotor circuit, there is no magnetic field created in the rotor to produce a torque.
9. Reconnect the wound rotor motor as shown in Figure 38-7. **Do not** disconnect the electrodynamometer or prime mover/dynamometer.
10. Set the rotor resistance control knob to the full clockwise position. This connects maximum resistance in the rotor circuit.

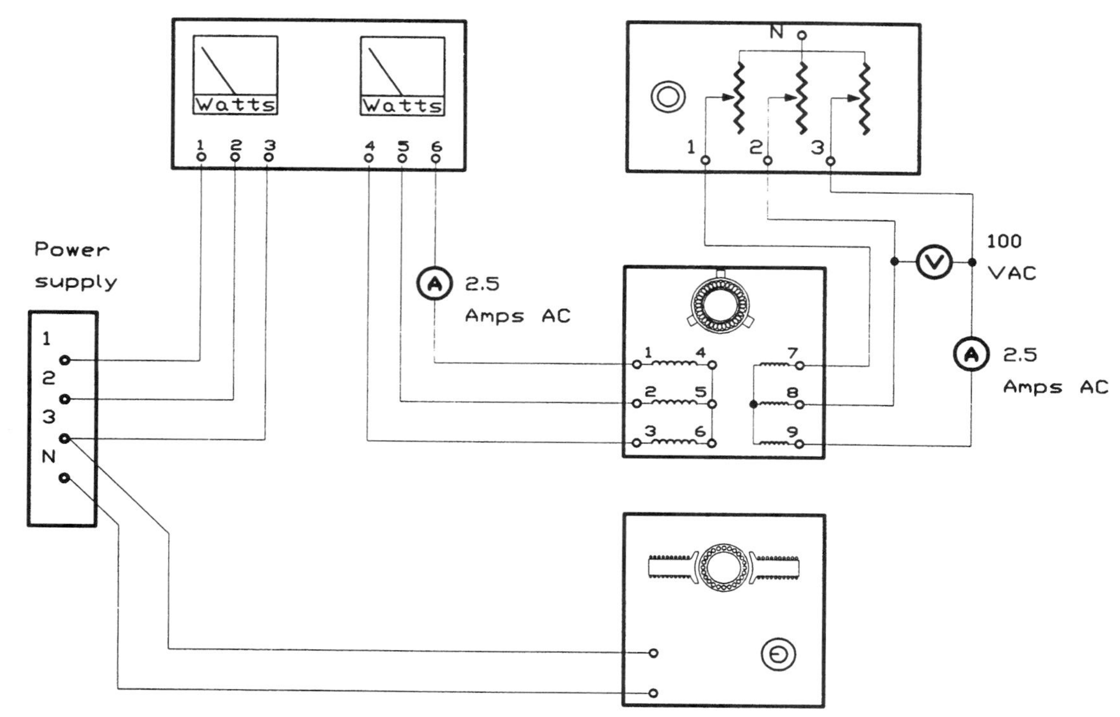

FIGURE 38-7 Wound rotor motor connection

11. Turn on the power supply and observe the direction of rotation of the motor. If the motor is not turning in the clockwise direction as you face the motor, turn off the power supply and change any two of the motor stator leads. Then turn the power supply back on and adjust the electrodynamometer for a torque of 8 lb.-in.

12. Measure and record the following values:

    I_{STATOR} = ______________ A

    I_{ROTOR} = ______________ A

    E_{ROTOR} = ______________ V

    RPM = ______________

    Power in = ______________ W

13. Compute the output horsepower using the formula

$$hp = \frac{6.28 \times RPM \times L \times P}{33,000}$$

where hp = horsepower
6.28 is a constant
RPM = motor speed in revolutions per minute
L = length (1/12 foot)
P = pounds of torque
33,000 is a constant

    hp = ______________ hp

14. Compute the output power by multiplying the output horsepower by 746.

    Power out = ______________ W

15. Compute the efficiency of the motor by using the formula

$$\text{eff.} = \frac{\text{power out}}{\text{power in}} \times 100$$

    Eff. = ______________ %

16. In this experiment, the resistance connected in the rotor circuit will be decreased in three steps. This will provide four different measurements of horsepower and efficiency. The voltage induced in the rotor circuit will be used as a guide in decreasing the rotor resistance. As rotor resistance is reduced and the rotor increases in speed, the amount of induced voltage will decrease. This reduction of voltage will be used to determine how much resistance to remove from the rotor circuit in each step. Divide the amount of rotor voltage measured in step 12 by 3.

    E_{STEP16} = ______________ V

17. Reduce the amount of rotor resistance until the voltage induced in the rotor decreases by the amount of voltage computed in step 16. Maintain a torque of 8 lb-in. on the electrodynamometer.

18. Measure and record the following values:

    I_{STATOR} = ______________ A

    I_{ROTOR} = ______________ A

    E_{ROTOR} = ______________ V

    RPM = ______________

    Power in = ______________ W

19. Compute the horsepower output of the motor.

    hp = ______________

20. Compute the output power.

    Power out = ______________ W

21. Compute the efficiency of the motor.

    Eff. = ______________ %

22. Decrease the rotor resistance until the induced voltage in the rotor drops by the amount computed in step 16. Maintain a torque of 8 lb-in. on the electrodynamometer.

23. Measure and record the following values:

    I_{STATOR} = ______________ A

    I_{ROTOR} = ______________ A

    E_{ROTOR} = ______________ V

    RPM = ______________

    Power in = ______________ W

24. Compute the horsepower output of the motor.

    hp = ______________ hp

25. Compute the output power.

    Power out = ______________ W

26. Compute the efficiency of the motor.

    Eff. = ______________ %

27. Decrease the rotor resistance until the control knob is in the full counterclockwise direction. At this point there should be zero resistance connected in the rotor circuit. Maintain a torque of 8 lb-in. on the electrodynamometer.

28. Measure and record the following values:

    I_{STATOR} = ______________ A

    I_{ROTOR} = ______________ A

    E_{ROTOR} = ______________ V

    RPM = ______________

    Power in = ______________ W

29. Compute the horsepower output of the motor.

    hp = ______________ hp

30. Compute the output power.

    Power out = ______________ W

31. Compute the efficiency of the motor.

    Eff. = ______________ %

32. **Turn off the power and disconnect the circuit.**

33. Return the components to their proper place.

Review Questions

1. What is the difference between a squirrel cage motor and a wound rotor motor?

   ____

2. What is the advantage of the wound rotor motor over the squirrel cage motor?

   ____

3. Name three factors that determine the amount of voltage induced in the rotor of a wound rotor motor.

   A. ____

   B. ____

   C. ____

4. Why will the rotor of a wound rotor motor not turn if the rotor circuit is left open with no resistance connected to it?

   ____

5. Why should power not be applied to the stator of a wound rotor motor if there is no resistance connected to the rotor circuit?

   ____

6. Why is the starting torque of a wound rotor motor higher per amp of starting current than that of a squirrel cage motor?

   ____

Exercise 39

The Synchronous Motor

Objectives

After completing this lab you should be able to:

- Discuss the operation of a synchronous motor.
- Explain why a synchronous motor operates at a constant speed.
- Explain the use of the amortisseur winding.
- Connect a synchronous motor and make measurements using test instruments.

Materials and Equipment

Synchronous machine module	EMS 8241
Power supply module	EMS 8821
DC metering module	EMS 8412
Three-phase wattmeter module	EMS 8441
Electrodynamometer module	EMS 8911
or prime mover/dynamometer	EMS 8960
Hand-held tachometer	EMS 8920

Discussion

The three-phase synchronous motor has several characteristics that separate it from the other types of three-phase motors. Some of these characteristics are:

1. The synchronous motor is not an induction motor. It does not depend on induced current in the rotor to produce a torque.
2. It will operate at a constant speed from full load to no load.
3. The synchronous motor must have DC excitation to operate.
4. It will operate at the speed of the rotating magnetic field (***synchronous speed***).
5. It has the ability to correct its own power factor and the power factor of other devices connected to the same line.

ROTOR CONSTRUCTION

The synchronous motor has the same type of stator windings as the other two three-phase motors. The rotor of a synchronous motor has windings similar to the wound rotor induction motor, as shown in Figure 39-1. Notice that the winding in the rotor of a synchronous motor is different, however. The winding of a synchronous motor is one continuous set of coils instead of three different sets as is the case with the wound rotor motor. Notice also that the synchronous motor has only two sliprings on its shaft as opposed to three on the shaft of a wound rotor motor.

Synchronous motors are also constructed with brushless exciters like those used on large alternators. The synchronous machine used in the Lab-Volt EMS training system uses two sliprings on the shaft to provide excitation for the rotor.

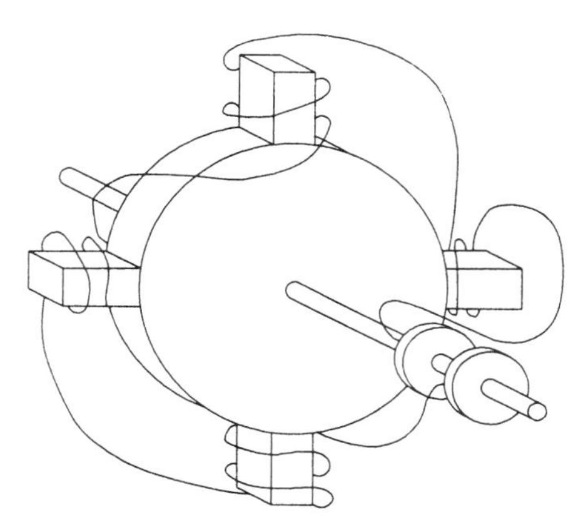

FIGURE 39-1 Synchronous motor rotor

STARTING A SYNCHRONOUS MOTOR

The rotor of a synchronous motor also contains a set of type A squirrel cage bars. This set of squirrel cage bars is used to start the motor and is known as the amortisseur winding, as shown in Figure 39-2. When power is first connected to the stator, the rotating magnetic field cuts through the type A squirrel cage bars. The cutting action of the field induces a current into the squirrel cage winding. The current flow through the amortisseur winding produces a rotor magnetic field that is attracted to the rotating magnetic field of the stator. This causes the rotor to begin turning in the direction of rotation of the stator field. When the rotor has accelerated to a speed that is close to the synchronous speed of the field, direct current is connected to the rotor through the sliprings on the rotor shaft, as in Figure 39-3.

When DC current is applied to the rotor, the windings of the rotor become electromagnets. The electromagnetic field of the rotor locks in step with the rotating magnetic field of the stator. The rotor will now turn at the same speed as the rotating magnetic field. When the rotor turns at the synchronous speed of the field, there is no more cutting action between the stator field and the amortisseur winding. This causes the current flow in the amortisseur winding to cease.

Notice that the synchronous motor starts as a squirrel cage induction motor. Because the rotor bars used are type A, they have a relatively high resistance, which gives the motor good starting torque and low starting current. A synchronous motor must never be started with DC current connected to the rotor. If DC current is applied to the rotor, the field poles of the rotor become electromagnets. When the stator is energized, the rotating magnetic field begins turning at synchronous speed. The electromagnets are alternately attracted and repelled by the stator field. As a result, the rotor does not turn.

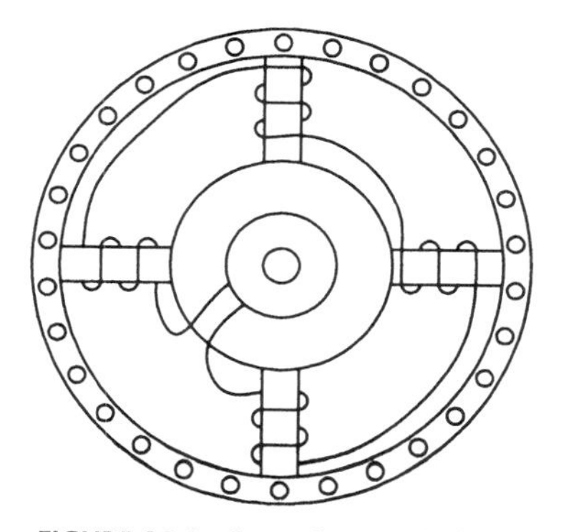

FIGURE 39-2 Amortisseur windings

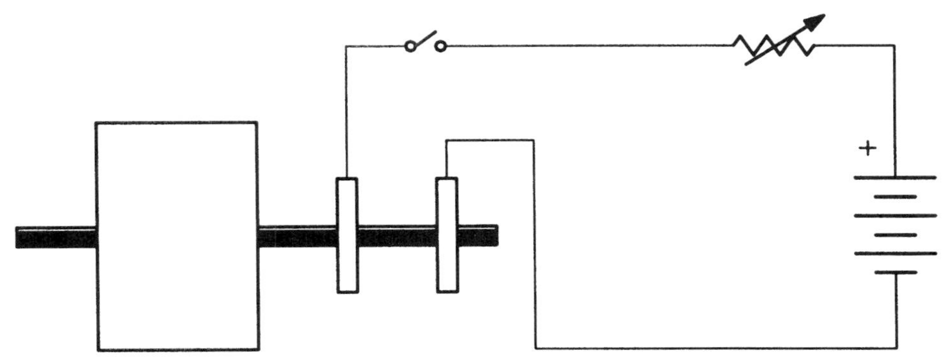

FIGURE 39-3 Direct current is applied to the rotor through the sliprings.

THE FIELD DISCHARGE RESISTOR

When the stator winding is first energized, the rotating magnetic field cuts through the rotor winding at a fast rate of speed. This causes a large amount of voltage to be induced into the winding of the rotor. To prevent this induced voltage from becoming excessive, a resistor is connected across the winding. This resistor is known as the field discharge resistor, shown in Figure 39-4. It also helps to reduce the voltage induced into the rotor by the collapsing magnetic field when the DC current is disconnected from the rotor.

CONSTANT SPEED OPERATION

Although the synchronous motor starts as an induction motor, it does not operate as one. After the amortisseur winding has been used to accelerate the rotor to about 95% of the speed of the rotating magnetic field, direct current is connected to the rotor and the electromagnets lock in step with the rotating field. Notice that the synchronous motor does not depend on induced voltage from the stator field to produce a magnetic field in the rotor. The magnetic field of the rotor is produced by external DC current applied to the rotor. This is the reason that the synchronous motor has the ability to operate at the speed of the rotating magnetic field. As load is added to the motor, the magnetic field of the rotor remains locked with the rotating magnetic field of the stator, and the rotor continues to turn at the same speed.

As load is added to the motor, the magnetic fields of the rotor and stator become stressed. This stress can be demonstrated with the Lab-Volt EMS system by marking the timing pulley on the shaft of the synchronous machine. If the electrodynamometer is used to load the motor, the stroboscope can be used to show that the synchronous

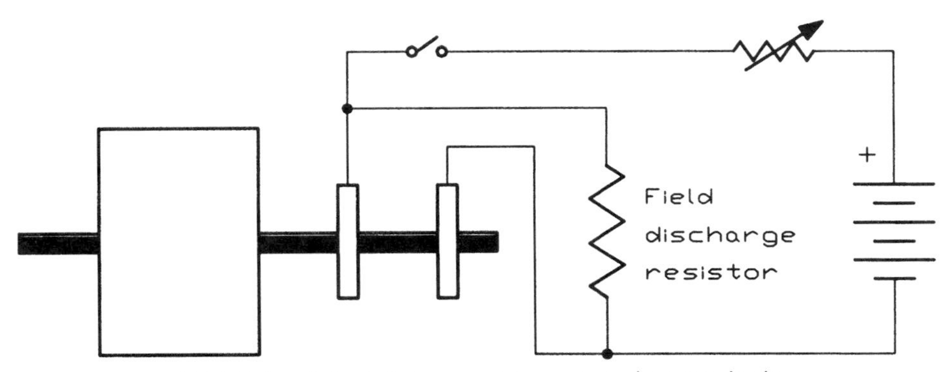

FIGURE 39-4 A field discharge resistor protects the rotor circuit.

motor maintains a constant speed even as load is added. It will be seen, however, that the mark will move in a direction opposite that of the direction of rotation as load is added to the motor. The amount of movement is proportional to the stress between the rotor and stator magnetic fields, and it increases with an increase of load. This stress is similar to connecting the north and south ends of two magnets together and then trying to pull them apart. If the force being used to pull the magnets apart becomes greater than the strength of the magnetic attraction, the magnetic coupling will be broken and the magnets can be separated. The same is true for the synchronous motor. If the load on the motor becomes too great, the rotor will be pulled out of synchronism with the rotating magnetic field. If this should happen, the motor must be stopped and restarted.

THE POWER SUPPLY

The DC power supply of a synchronous motor can be provided by several methods. The most common of these methods is either a small DC generator mounted to the shaft of the motor or an electronic power supply that converts the AC line voltage into DC voltage.

Name ____ Date ________________

Procedure

1. Remove the EMS 8241 synchronous machine module from the mobile console.
2. Looking at the back of the motor, how many brushes and sliprings are connected to the shaft?

   ____

3. Observe the rotor winding. This winding is used to produce electromagnets in the rotor when the DC excitation current is applied.
4. Replace the EMS 8241 synchronous machine module in the mobile console.
5. Looking at the front panel of the module, locate the stator winding connections. Notice that these windings are labeled 1–4, 2–5, and 3–6. What are the voltage and current ratings of these windings?

   ____________ V ____________ A

6. Locate the rotor connection leads. These terminals are labeled 7 and 8. What is the rated input voltage of the rotor?

   ____________ VDC

7. Locate the switch on the front panel. This switch is used to connect DC voltage to the rotor after the motor is running. It is known as the excitation switch.

NOTE: The synchronous machine should never be started with this switch in the closed (on) position.

8. Locate the variable resistor connected in series with the rotor circuit. This resistor is used to control the amount of excitation current in the rotor. What is the maximum resistance of this variable resistor?

   ____________ Ω

9. Replace the motor in the cabinet and connect the synchronous machine to the dynamometer with the timing belt.
10. Connect the circuit shown in Figure 39-5 if using the electrodynamometer. If the prime mover/dynamometer is being used, connect the 24 VAC low-voltage connection to the power supply.
11. Set the excitation control rheostat located on the front panel of the synchronous machine to the full counterclockwise position.
12. Set the dynamometer control knob to the full counterclockwise position.
13. Turn on the power supply and check the direction of rotation of the motor. The motor should turn in a clockwise direction as you face the motor. **If it does not, turn off the power supply** and change any two of the stator leads. Then turn the power supply back on.
14. Close the excitation switch and adjust the control rheostat until a current of 0.7 ADC flows through the rotor circuit.
15. Fill in the missing values in the chart shown in Figure 39-6. This can be done by reading instruments and using the formulas shown.

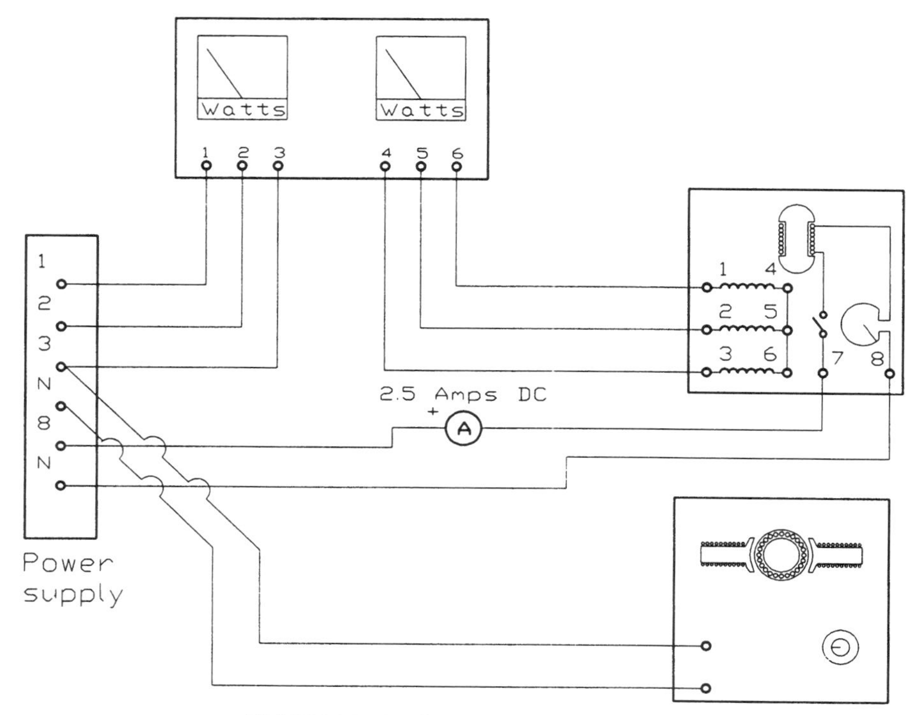

FIGURE 39-5 Synchronous motor connection

The speed and input power are found by using test instruments. The output horsepower can be found by using the formula

$$hp = \frac{6.28 \times RPM \times L \times P}{33,000}$$

where 6.28 is a constant
RPM = speed in revolutions per minute
L = length in feet (the torque is in lb-in., so this will be 0.0833)
P = pounds of force
33,000 is a constant

The output power can be computed by using the formula

$$\text{output power} = \text{output horsepower} \times 746$$

The efficiency can be computed by using the formula

$$\text{eff.} = \frac{\text{output power}}{\text{input power}} \times 100$$

16. **Turn off the power supply and disconnect the circuit.**
17. Return the components to their proper place.

Load torque (lb-in.)	Speed (RPM)	Input power (watts)	Output horsepower (hp)	Output power (watts)	Eff. (%)
0					
2					
4					
6					
8					
10					
12					

FIGURE 39-6

Review Questions

1. Is the synchronous motor an induction motor?

 ____

2. What is the amortisseur winding?

 ____

3. Why must a synchronous motor never be started when DC excitation current is applied to the rotor?

 ____

4. Name three characteristics that make the synchronous motor different from an induction motor.

 A. ____

 B. ____

 C. ____

5. What is the function of the field discharge resistor?

 ____

Exercise 40

The Synchronous Condenser

Objectives

After completing this lab you should be able to:

- Discuss how a synchronous motor can be used to correct power factor.
- Connect a synchronous motor and determine power factor using test instruments.
- Correct the power factor with a synchronous motor.

Materials and Equipment

Synchronous machine module	EMS 8241
Power supply module	EMS 8821
Variable-resistance module	EMS 8311
Variable-inductance module	EMS 8321
Three-phase wattmeter module	EMS 8441
DC metering module	EMS 8412

Discussion

The synchronous motor has the ability to correct its own power factor and the power factor of other devices connected to the same line. Synchronous motors are sometimes operated at no load and are used for the specific purpose of power factor correction. When this is done, they are generally referred to as synchronous condensers instead of synchronous motors.

CONTROLLING POWER FACTOR CORRECTION

The amount of power factor correction is controlled by the amount of excitation current in the rotor. If the rotor of a synchronous motor is underexcited, the motor will have a lagging power factor like a common induction motor. As rotor excitation current is increased, the synchronous motor appears to be more capacitive. When the excitation current reaches a point where the power factor of the motor is at unity, or 100%, it is at the normal excitation level. If the excitation current is increased above the normal level, the motor will have a leading power factor and appear as a capacitive load. Because capacitance has now been added to the line, it will correct the lagging power factor of other inductive devices connected to the same line.

ADVANTAGES OF THE SYNCHRONOUS CONDENSER

The advantage of using a synchronous condenser over a bank of capacitors for power factor correction is that the amount of correction is easily controlled. When a bank of capacitors is used for correcting power factor, capacitors must be added to or removed from the bank if a change in the amount of correction is needed. When a synchronous condenser is used, only the excitation current must be changed to cause an alteration of power factor.

Name ______________________________ Date ______________

Procedure

In this experiment, the resistive and inductive load modules will be used to simulate other plant loads. The inductive load bank will cause a lagging power factor similar to that of other motors being connected to the same line. The synchronous machine will be operated at no load in this experiment. Its sole purpose in the circuit will be to correct the power factor.

1. Connect the circuit shown in Figure 40-1.
2. Set the excitation rheostat located on the front of the synchronous machine module to the full counterclockwise position.
3. Set the inductive load module for a value of 1.6 H per phase. (Close each of the three switches to the 1.6-H inductors.)
4. Set the resistive load module for a value of 300 Ω per phase. (Close the three 300-Ω switches.)
5. Turn on the power supply and close the excitation switch located on the front of the synchronous machine module.
6. Measure the amount of excitation current.

   I = ____________ ADC

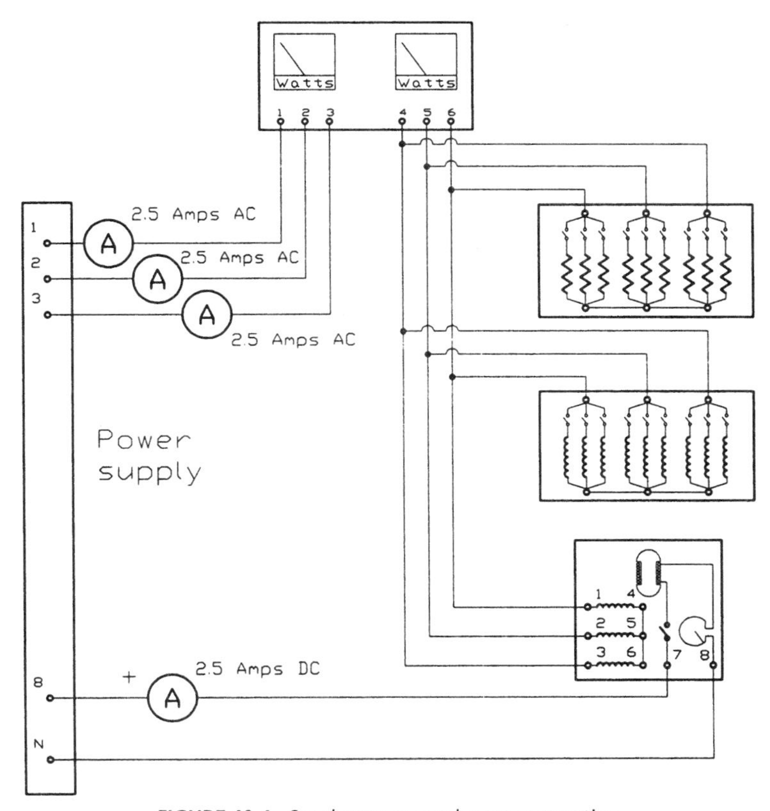

FIGURE 40-1 Synchronous condenser connection

7. Measure the amount of line current in the circuit.

   I = ______________ AAC

8. Measure the amount of true power in the circuit.

   P = ______________ W

9. Compute the apparent power in the circuit using the formula

$$VA = 1.732 \times E_{LINE} \times I_{LINE}$$

   VA = ______________

10. Compute the power factor using the formula

$$PF = \frac{W}{VA} \times 100$$

   PF = ______________ %

11. Slowly increase the amount of excitation current to the rotor of the synchronous machine while observing the AC ammeters connected in series with the line. Continue to increase the excitation current until the ammeters drop to their lowest level. At this point, the power factor should be corrected.

12. Measure the amount of excitation current.

   I = ______________ ADC

13. Measure the amount of line current.

   I = ______________ AAC

14. Measure the true power in the circuit.

   P = ______________ W

15. Compute the amount of apparent power in the circuit.

   VA = ______________

16. Compute the power factor.

   PF = ______________ %

17. Compare this power factor value with the previous power factor value.
18. Turn off the power supply and open the excitation switch.
19. Disconnect the circuit and return the components to their proper place.

Review Questions

1. When is a synchronous motor a synchronous condenser?

   ____

2. What determines when a synchronous motor is at normal excitation?

   ____

3. How can a synchronous motor be made to have a leading power factor?

   ____

4. Is the excitation current AC or DC?

   ____

Exercise 41

Split-Phase Motors: The Resistance Start Induction Run

Objectives

After completing this lab you should be able to:

- Discuss the operation of split-phase motors.
- Connect a resistance start induction run motor and make measurements with test instruments.
- Reverse the direction of rotation of a split-phase motor.
- Discuss the starting characteristics of a resistance start induction run motor.

Materials and Equipment

AC ammeter module	EMS 8425
Power supply module	EMS 8821
Electrodynamometer module	EMS 8911
or prime mover/dynamometer	EMS 8960
Hand-held tachometer	EMS 8920
Single-phase wattmeter module	EMS 8431
Capacitor start motor module	EMS 8251
Ohmmeter (supplied by student)	

Discussion

Split-phase motors fall into three general classifications:

1. The resistance start induction run motor
2. The capacitor start induction run motor
3. The capacitor start capacitor run motor

Although all of these motors have different operating characteristics, they are similar in construction. Split-phase motors derive their name from the manner in which they operate. Recall from the study of three-phase motors that the basic principle of operation of an AC induction motor is that of a rotating magnetic field. One of the factors that causes the field to rotate is the fact that the three line voltages are 120° out of phase with each other. In a single-phase power system, there is no other phase that can be used to produce a rotating field.

THE TWO-PHASE SYSTEM

In some parts of the world, two-phase power is produced. A two-phase system is produced by having an alternator with two sets of coils wound 90° out of phase with each other, as shown in Figure 41-1. The voltages of a two-phase system are, therefore, 90° out of phase with each other, as shown in Figure 41-2. The two out-of-phase voltages can be used to produce a rotating magnetic field. Because there have to be two voltages

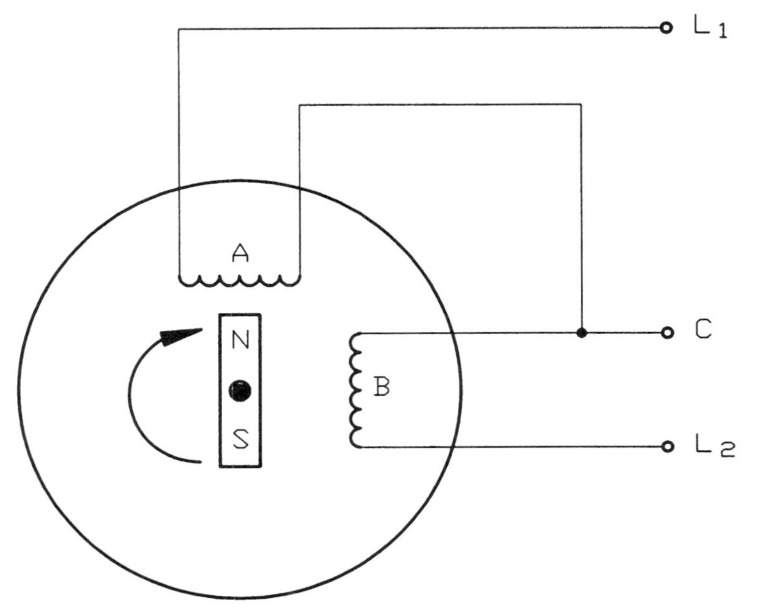

FIGURE 41-1 Two-phase alternator

or currents out of phase with each other to produce a rotating magnetic field, split-phase motors use two separate windings to create a phase difference between the currents in each of these windings. These motors literally "split" one phase and produce a second phase, hence the name split-phase motor.

STATOR WINDINGS

The stator of a split-phase motor contains two separate windings. The start winding and the run winding. The start winding is made of thin small wire and is placed near the top of the stator core. The run winding is made of relatively thick large wire and is placed in the bottom of the stator core.

The fact that the start winding is made from small wire and placed near the top of the stator core causes it to have a higher resistance than the run winding. These two windings are connected in parallel with each other, as shown in Figure 41-3. Since the

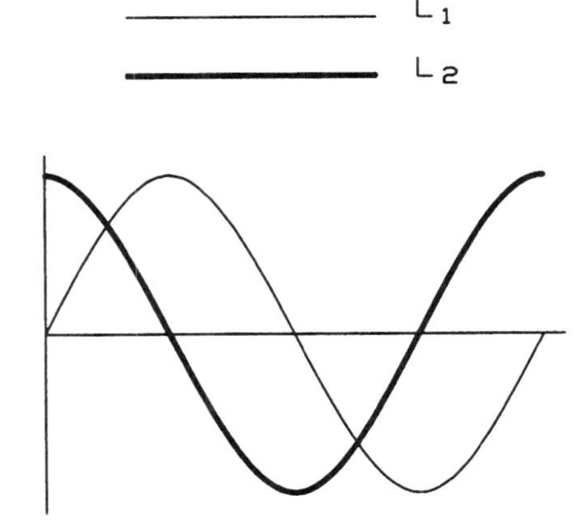

FIGURE 41-2 Two-phase voltages are 90° out of phase with each other.

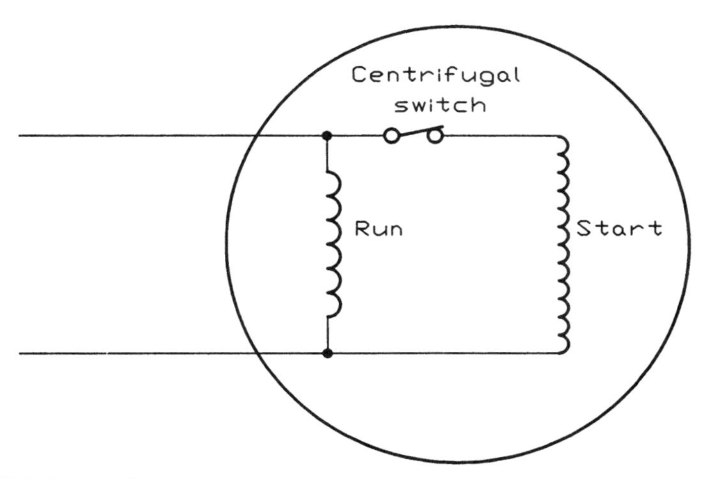

FIGURE 41-3 The start winding is connected in parallel with the run winding.

run winding is made from large wire and placed near the bottom of the stator core, it is more inductive than the start winding. This causes the current flowing through the run winding to be out of phase with the current in the start winding. These two out-of-phase currents produce the rotating magnetic field in the stator. The speed of the rotating magnetic field is determined by the number of stator poles and the frequency.

Once the squirrel cage rotor has accelerated to a point that it is operating at about 75% of the rotating field speed, the start winding can be disconnected from the circuit. The rotor will continue to rotate because of the changing magnetic polarities of the run winding.

REVERSING DIRECTION OF ROTATION

The direction of rotation of a split-phase motor can be reversed by changing the start winding leads or the run winding leads, but not both. The rotation is generally reversed by changing the start winding leads with respect to the run winding. Some motor manufacturers bring both start winding leads to the outside of the motor. This permits the direction of rotation to be determined when the motor is installed.

DISCONNECTING THE START WINDING

As stated previously, when the rotor of a split-phase motor reaches about 75% of the speed of the rotating magnetic field, the start windings can be disconnected from the circuit. Open-case motors generally use a centrifugal switch mechanism to perform this job. A simple centrifugal switch mechanism is shown in Figure 41-4. This diagram is intended to illustrate the principle of operation of the switch. When the shaft is not turning, the bottom ring rides against the movable contact arm. The weight of the metal balls overcomes the spring tension pushing upward against the movable contact arm, and the contact is held closed. When the shaft turns, centrifugal force causes the metal balls to spin outward. As the metal balls spin outward, the bottom ring is mechanically lifted away from the movable contact. This permits the spring to open the contact.

It should be noted that not all start windings are disconnected with a centrifugal switch. This is especially true in air conditioners and refrigerators, where motors are often hermetically sealed. In these instances, starting relays are used to disconnect the start winding from the circuit at the proper time.

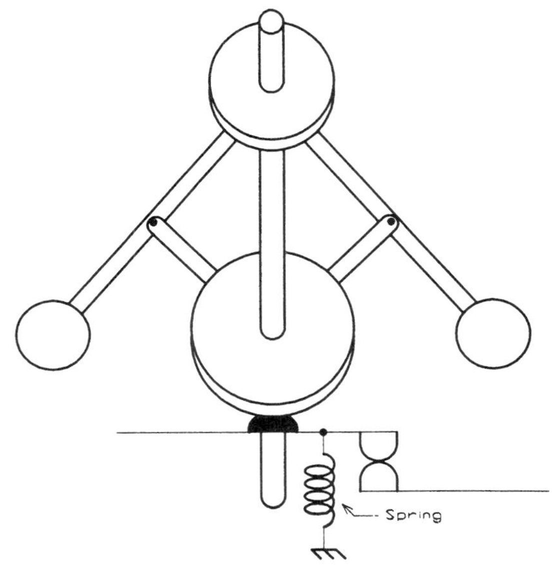

FIGURE 41-4 Basic centrifugal switch

DUAL-VOLTAGE MOTORS

Single-phase motors can also be constructed to operate on two separate voltages. These motors are designed to be connected to 120 or 240 V. A common connection for this type of motor contains two run windings and one start winding, as shown in Figure 41-5. The run windings are labeled T_1-T_2 and T_3-T_4. The start winding is labeled T_5 and T_6. In the circuit shown in Figure 41-5, the windings have been connected for operation on a 240-V line. Each winding is rated at 120 V. The two run windings are connected in series, which causes each to have a voltage drop of 120 V when connected to 240 V. The start winding is connected in parallel with one of the run windings. This causes the start winding to have an applied voltage of 120 V also. Notice that each of the windings will have a voltage drop of 120 V, which is the rated voltage of each winding.

If the motor is to be operated on a 120-V line, the windings are connected in parallel, as shown in Figure 41-6. Because these windings are connected in parallel, each will have 120 V applied to it.

Although it is common for split-phase motors designed to operate on dual voltage to have only one start winding, some motors will contain two start windings as well as two run windings. When this is the case, the second start winding is labeled T_7 and T_8. When the motor is to be operated on 120 V, the second start winding will be connected in parallel with the first. If the motor is to be operated on 240 V, the two start windings will be connected in series. Since it is possible for motors to have start windings numbered from T_5 to T_8, it is not uncommon to find motors with the start winding leads labeled T_5 and T_8 although there is only one start winding instead of two.

MOTOR POWER CONSUMPTION

It should be noted that the motor does not use less energy when connected to 240 V than it does when connected to 120 V. Power is measured in watts, and the watts will be the same regardless of the connection. When the motor is connected to operate on 240 V, it will have half the current draw as it does on a 120-V connection. Therefore, the amount of power used is the same. For example, assume that a motor has a current

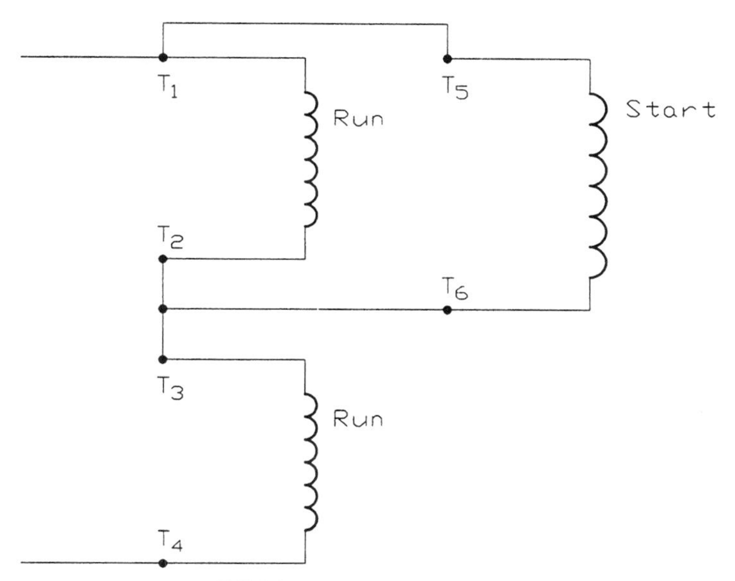

FIGURE 41-5 240-V connection

draw of 5 A when connected to 240 V and 10 A when connected to 120 V. Watts can be computed by multiplying the voltage by the current. When the motor is connected to 240 V, the amount of power used is 1200 W (240 V × 5 A = 1200 W). When the motor is connected to 120 V, the amount of power used is 1200 W (120 V × 10 A = 1200 W).

The 240-V connection is generally preferred, however, because the lower current draw causes less voltage drop on the line supplying power to the motor. If the motor is located a long distance from the power source, the voltage drop of the wire can become very important to the operation of the motor.

MOTOR STARTING

The resistance start induction run motor develops a rotating magnetic field by using the resistance of the start winding. Because the start winding has a higher resistance than the run winding, the current in the start winding will be out of phase with the current in the run winding. Maximum starting torque is developed when these two currents are

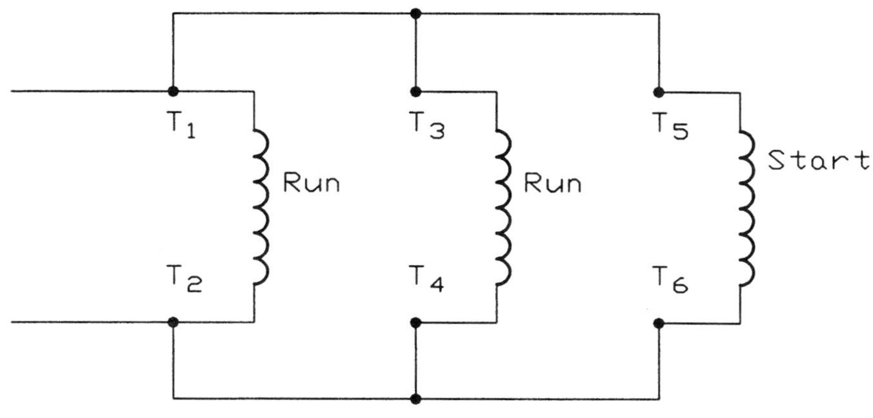

FIGURE 41-6 120-V connection

90° out of phase with each other. The run winding, however, is not a pure inductor. It has some resistance in the wire used to make it. This causes the current flowing in the run winding to be less than 90° out of phase with the voltage.

The start winding is not a pure resistance. This winding has some inductance, which causes the current to lag the voltage by some amount. As a result, there is not a 90°-phase difference between the current in the start winding and the current in the run winding, as shown in Figure 41-7. The actual phase angle difference between these two currents in this type of motor is about 40°. This causes the resistance start motor to have a relatively poor starting torque.

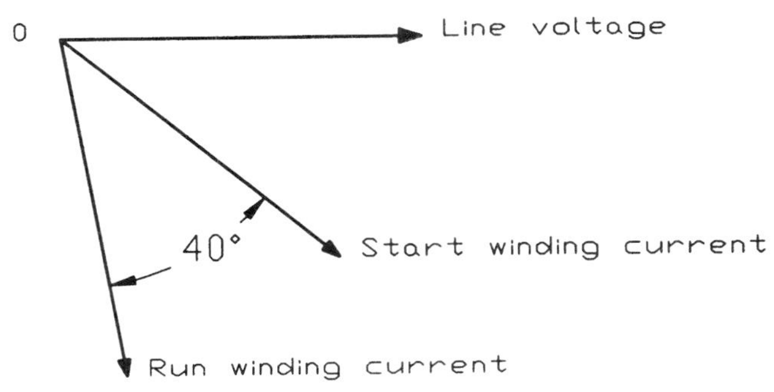

FIGURE 41-7 Start winding and run winding currents are about 40° out of phase with each other.

Name ______________________________ Date ______________

Procedure

1. Remove the capacitor start motor module Figure 41-8 from the mobile console.
2. Examine the rear of the module. Locate the centrifugal switch mechanism. Notice that it consists of two parts. One part is the rotating member and is attached to the motor shaft. This part contains the springs, weights, and sliding ring. The second part is stationary and contains the spring-loaded contact connected in series with the start winding.
3. Locate the run and start windings on the stator. The run winding is made with the larger wire and is placed near the bottom of the stator slots. The start winding is made with the smaller wire and is placed near the top of the stator slots.
4. How many stator poles does this motor contain?

   ____

5. Replace the motor module in the mobile console.
6. Using an ohmmeter, measure the resistance of the run winding between points 1 and 2 on the front of the motor module.

   ______________ Ω

7. Measure the resistance of the start winding between points 3 and 4.

   ______________ Ω

8. Connect the circuit shown in Figure 41-9 if using the electrodynamometer. If the prime mover/dynamometer is being used, connect the 24 VAC low-voltage connection to the power supply. Notice that the capacitor has **not** been connected in the motor circuit. This permits the motor to operate as a resistance start, induction-run motor instead of a capacitor-start, induction-run motor.
9. Connect the capacitor start motor to the electrodynamometer with the timing belt, and set the control knob located on the electrodynamometer module to the full counterclockwise position.

FIGURE 41-8 Capacitor start motor. (Courtesy of Lab-Volt® Systems, Inc.)

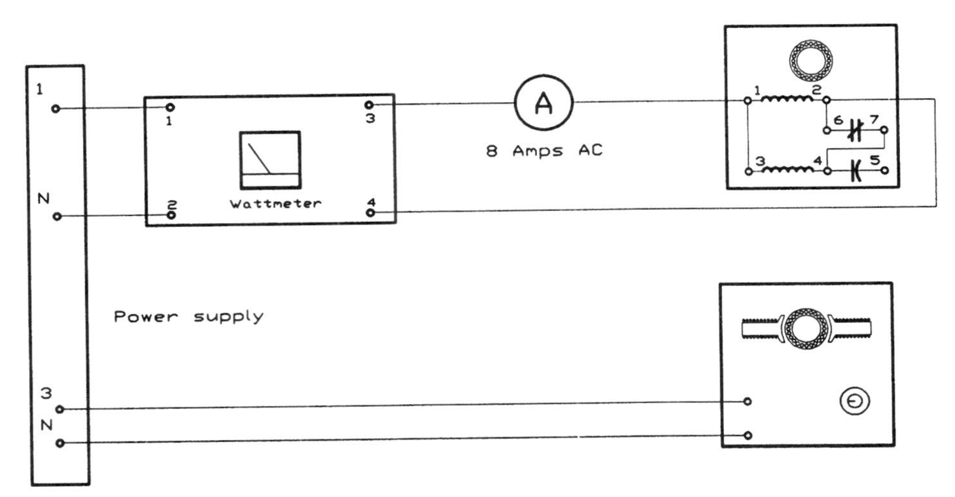

FIGURE 41-9 Resistance start induction run motor connection

10. Turn on the power supply and notice the direction of rotation. What is the direction of rotation?

    ____

11. **Turn off the power supply.**
12. Reverse the connection to the start winding by changing the wires connected to terminals 3 and 4 on the motor module.
13. Turn on the power supply and note the direction of rotation. Does the motor turn in the clockwise or counterclockwise direction as you face the motor?

    ____

14. **Turn off the power supply** and set the start winding connection so the motor turns in the clockwise direction as you face the motor.
15. Fill in the chart shown in Figure 41-10 by measuring values with test instruments and computing values with the following formulas.

    The horsepower can be computed using the formula

$$hp = \frac{6.28 \times RPM \times L \times P}{33,000}$$

where 6.28 is a constant
RPM = speed in revolutions per minute
L = length in feet (0.08333)
P = pounds of force
33,000 is a constant

The output power can be computed by using the formula

$$\text{output power} = \text{output horsepower} \times 746$$

The efficiency can be computed by using the formula

$$\text{eff.} = \frac{\text{output power}}{\text{input power}} \times 100$$

Load torque (lb-in.)	Armature current (amps)	Speed (RPM)	Input power (watts)	Output horsepower (hp)	Output power (watts)	Eff. %
0						
2						
4						
6						
8						
10						

FIGURE 41-10

16. **Turn off the power supply and disconnect the circuit.**
17. Return the components to their proper place.

Review Questions

1. What is a split-phase motor?

   ____

2. What are the three basic types of split-phase motor?

   A. ______________________________

   B. ______________________________

   C. ______________________________

3. Explain the difference in construction of run windings and start windings.

   Run winding ______________________________

   Start winding ______________________________

4. How many degrees out of phase should the current in the start winding be with the current in the run winding to develop maximum starting torque?

   ____________°

5. What is the centrifugal switch used for?

   ____

6. How many degrees out of phase are the voltages of a two-phase system?

   ____________°

7. How can the direction of rotation of a split-phase motor be reversed?

   ____

Exercise 42

Split-Phase Motors: The Capacitor Start Induction Run

Objectives

After completing this lab you should be able to:

- Discuss the differences between resistance start induction run and capacitor start induction run motors.
- State the reason that a capacitor start motor has a higher starting torque than a resistance start motor.
- Connect a capacitor start induction run motor and make measurements using test instruments.

Materials and Equipment

AC ammeter module	EMS 8425
Power supply module	EMS 8821
Electrodynamometer module	EMS 8911
or prime mover/dynamometer	EMS 8960
Hand-held tachometer	EMS 8920
Single-phase wattmeter module	EMS 8431
Capacitor start motor module	EMS 8251

Discussion

The capacitor start induction run motor has the same running characteristics as the resistance start induction run motor. The capacitor start motor has improved starting torque, however. The capacitor is connected in series with the start winding, as shown in Figure 42-1. Because this capacitor is in the circuit for only the amount of time needed to start the motor, an AC electrolytic capacitor is generally used as the starting capacitor for this motor. This permits a large amount of capacitance in a small case size, but limits the times this motor should be started over a short period of time. Most manufacturers do not recommend starting this type of motor more than eight times per hour, or the starting capacitor could be damaged.

STARTING CHARACTERISTICS

The capacitor causes an increase in starting torque by shifting the phase angle of the current flowing in the start winding. The start winding current is now capacitive, which causes the current to lead the voltage. This produces a greater phase angle difference between the run and start winding currents, as shown in Figure 42-2. If the correct amount of capacitance is used, a phase angle difference of 90° can be developed between these two currents, which results in maximum starting torque. If the amount of capacitance is too great, the start winding current will become greater than 90° out of phase with the run winding current, and the starting torque will decrease. When the

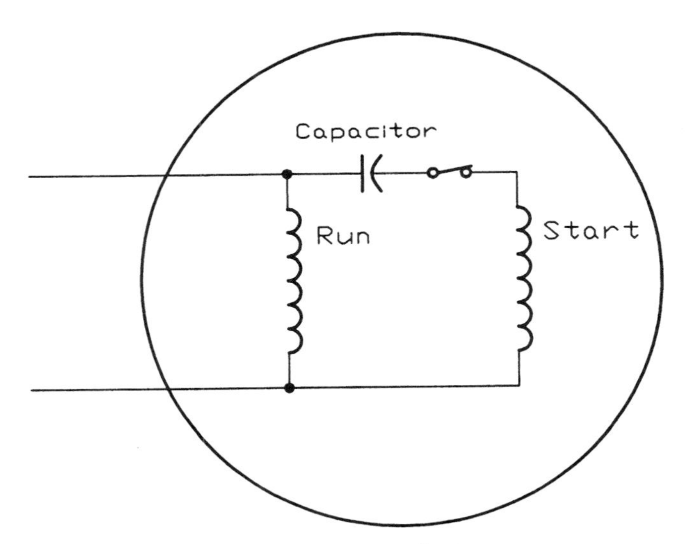

FIGURE 42-1 A capacitor start induction run motor

starting capacitor for this type of motor is replaced, the microfarad rating recommended by the manufacturer should be used. If necessary, a capacitor of higher voltage rating can be used, but a capacitor of lower voltage rating should never be used.

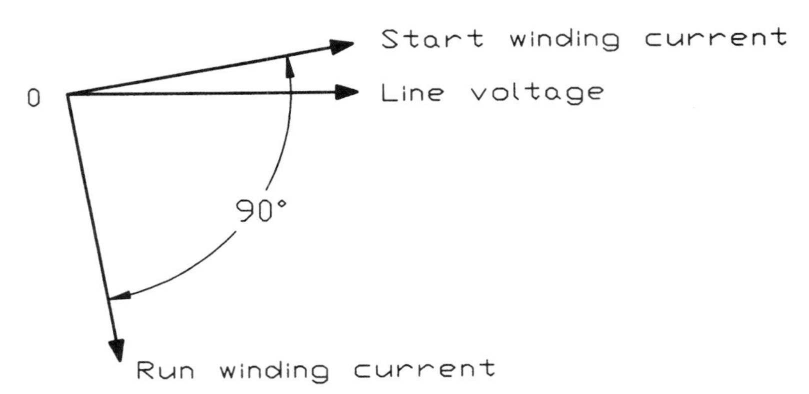

FIGURE 42-2 Start winding and run winding currents are 90° out of phase with each other.

Name ____ Date ________________

Procedure

1. Connect the circuit shown in Figure 42-3 if using the electrodynamometer. If the prime mover/dynamometer is being used, connect the 24 VAC low-voltage connection to the power supply.
2. Connect the capacitor start motor to the dynamometer with the timing belt, and set the control knob to the full counterclockwise position.
3. Turn on the power supply and notice the direction of rotation. Does the motor rotate in the clockwise or counterclockwise direction?

   ____

4. **Turn off the power supply.**
5. Reverse the connection to the start winding by changing the wires connected to terminals 3 and 5 on the motor module.
6. Turn on the power supply and note the direction of rotation. Does the motor turn in the clockwise or counterclockwise direction? Notice that the direction of rotation for the capacitor start motor is changed in the same way that it is for a resistance start motor.

   ____

7. **Turn off the power supply** and set the start winding connection so the motor turns in the clockwise direction as you face the motor.
8. Fill in the chart shown in Figure 42-4 by measuring values with test instruments and computing values with the following formulas.

   The horsepower can be computed using the formula

$$hp = \frac{6.28 \times RPM \times \times L \quad P}{33,000}$$

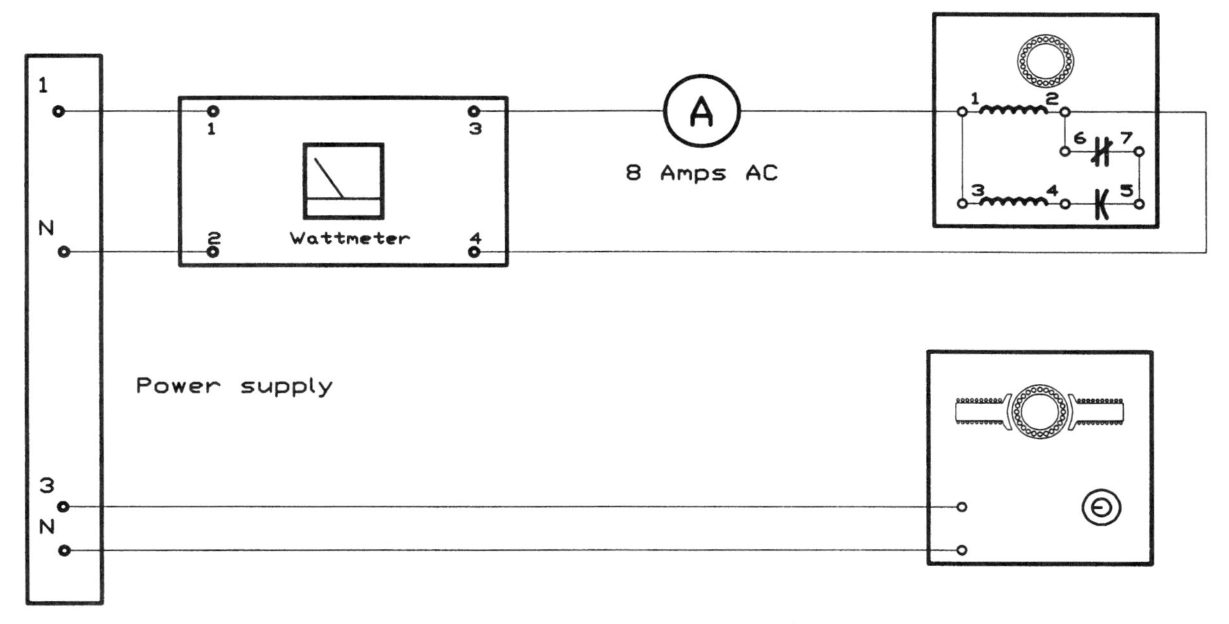

FIGURE 42-3 Capacitor start induction run motor connection

Load torque (lb-in.)	Armature current (amps)	Speed (RPM)	Input power (watts)	Output horsepower (hp)	Output power (watts)	Eff. %
0						
2						
4						
6						
8						
10						

FIGURE 42-4

where 6.28 is a constant
RPM = speed in revolutions per minute
L = length in feet
P = pounds of force
33,000 is a constant

The output power can be computed by using the formula

$$\text{output power} = \text{output horsepower} \times 746$$

The efficiency can be computed by using the formula

$$\text{eff.} = \frac{\text{output power}}{\text{input power}} \times 100$$

9. **Turn off the power supply and disconnect the circuit.**
10. Return the components to their proper place.

Review Questions

1. What type of capacitor is generally used as the starting capacitor for this type of motor?

____

2. Why does the capacitor improve the starting torque over a resistance start induction run motor?

____

3. Can the microfarad value of the starting capacitor be increased to improve the starting torque?

____

4. Explain the answer in question 3.

____

5. Is it possible to use a starting capacitor with a higher voltage rating without affecting the amount of starting torque produced by the motor?

____

Exercise 43

Power Factor Correction For Single-Phase Motors

Objectives

After completing this lab you should be able to:

- Discuss power factor correction for single-phase motors.
- Determine the power factor of a single-phase motor using test instruments.
- Compute the amount of capacitance needed to correct the power factor of a motor.
- Correct motor power factor by adding capacitance to the circuit.

Materials and Equipment

Power supply module	EMS 8821
AC ammeter module	EMS 8425
AC voltmeter module	EMS 8426
Electrodynamometer module	EMS 8911
or prime mover/dynamometer	EMS 8960
Single-phase wattmeter module	EMS 8431
Capacitor start motor module	EMS 8251
Variable-capacitance module	EMS 8331

Discussion

The power factor of a single-phase motor can be corrected just as the power factor of a three-phase motor can. Power factor is corrected by connecting capacitance in parallel with the motor circuit.

In this laboratory exercise, the power factor of a single-phase motor will be determined using test instruments. The amount of capacitance needed to correct the motor power factor will then be computed and connected to the circuit. In this experiment, however, it may be necessary to connect two capacitance load modules in parallel with each other. When all the capacitors in the EMS 8331 variable-capacitance module are connected in parallel, a total capacitance of 46.2 μF can be obtained. If this amount of capacitance is not enough, it will be necessary to connect another capacitance module if it is available. If another capacitance load module is not available, all of the available capacitance will be connected in the circuit. The power factor will then be recomputed for the circuit.

Name ______________________________ Date ______________

Procedure

1. Connect the circuit shown in Figure 43-1 if using the electrodynamometer. If the prime mover/dynamometer is being used, connect the 24 VAC low-voltage connection to the power supply.
2. Connect the motor to the dynamometer with the timing belt and set the dynamometer control knob to the full counter-clockwise direction.
3. Open all the switches on the variable-capacitance module.
4. Turn on the power supply and observe the direction of rotation of the motor. The motor should turn in the clockwise direction as you face the motor. **If it does not, turn off the power supply** and interchange the start winding leads. Then turn the power supply back on.
5. Adjust the electrodynamometer for a load of 8 lb-in. of torque.
6. Measure the amount of true power in the circuit with the wattmeter.

   ____________ W

7. Measure the amount of voltage across the motor and the amount of current flow through the motor.

   E = ____________ V I = ____________ A

8. Compute the apparent power in the circuit using the formula

$$VA = E \times I$$

   VA = ____________

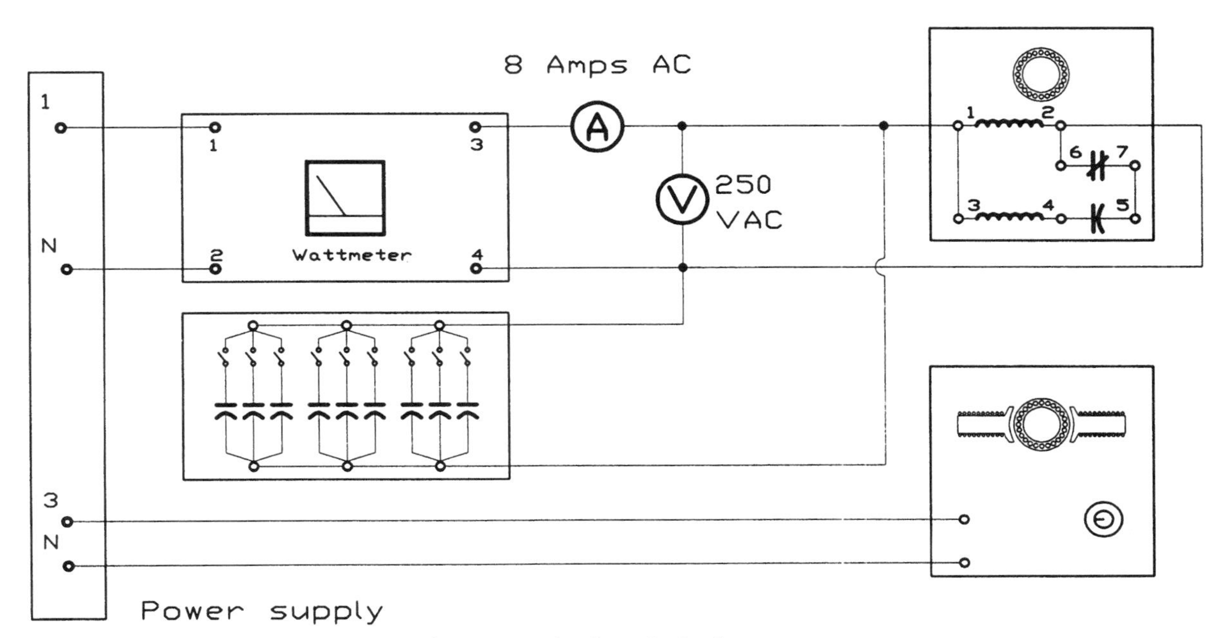

FIGURE 43-1 Power factor correction for a single-phase motor

9. Compute the circuit power factor using the formula

$$PF = \frac{W}{VA} \times 100$$

PF = ______________ %

10. Compute the amount of reactive power in the circuit using the formula

$$VARs = \sqrt{VA^2 - W^2}$$

VARs = ______________

11. Power factor is normally corrected to some value less than 100%. In this example, the power factor will be corrected to 95%. To find the amount of apparent power (volt-amps) needed to provide a power factor of 95%, divide the true power by 0.95.

$$VA = \frac{W}{0.95}$$

VA_{NEEDED} = ______________

12. Now that the apparent power needed to produce the desired power factor correction is known, the amount of reactive power needed to produce this apparent power can be computed. In the formula shown below, use the "needed" value of apparent power to compute the "needed" value of reactive power (VARs).

$$VARs_{NEEDED} = \sqrt{VA^2 - W^2}$$

$VARs_{NEEDED}$ = ______________

13. The amount of capacitive VAR that should be added to the circuit can be found by subtracting the present reactive power found in Step 10 from the amount of reactive power needed.

$$VARs_{CAPACITIVE} = VARs_{PRESENT} - VARs_{NEEDED}$$

$VARs_{CAPACITIVE}$ = ______________

14. Now that the amount of capacitive VARs needed in the circuit is known, this value can be used to compute the amount of capacitive current necessary to correct the power factor. The amount of capacitive current needed can be computed by using the formula

$$I_C = \frac{VARs}{E}$$

I_C = ______________ A

15. The capacitive current value can be used to calculate the amount of capacitive reactance needed in the circuit. To compute the capacitive reactance needed use the formula

$$X_C = \frac{E}{I_C}$$

X_C = ______________ Ω

16. The capacitive reactance value can be used to compute the amount of capacitance needed to correct the power factor by using the formula

$$C = \frac{1}{2\pi f X_C}$$

C = ______________ µF

NOTE: The formula gives the capacitance value in farads. This value must be converted to microfarads.

17. If possible, connect the amount of capacitance found in Step 16 in parallel with the motor by closing the proper switches on the variable-capacitance module. Observe the action of the AC ammeter while doing this.

18. Measure the amount of true power in the circuit.

    ______________ W

19. Measure the amount of voltage across the motor and the amount of circuit current.

    E = ______________ V I = ______________ A

20. Compute the circuit power factor.

    PF = ______________ %

21. **Turn off the power supply and disconnect the circuit.**

22. Return the components to their proper place.

Review Questions

1. Assume that a single-phase motor is connected to a 240-VAC line and has a current draw of 6 A. A wattmeter connected in the circuit indicates a true power value of 864 W. Compute the following values:

   VA = ______________

   PF = ______________ %

   VARs = ______________

2. Using the above values, compute the amount of capacitive current needed in the circuit to correct the power factor to 90%.

   I_C = ______________ A

3. Compute the amount of capacitive reactance needed to correct the power factor.

   X_C = ______________ Ω

4. Compute the amount of capacitance needed to correct the power factor to 90%.

   C = ______________ µF

Exercise 44

Split-Phase Motors: The Capacitor Start Capacitor Run

Objectives

After completing this lab you should be able to:

- Discuss the operation of a capacitor start capacitor run motor.
- Connect a capacitor start capacitor run motor and make measurements using test instruments.
- Determine operating characteristics and power factor from the measurements taken.

Materials and Equipment

Power supply module	EMS 8821
AC ammeter module	EMS 8425
AC voltmeter module	EMS 8426
Electrodynamometer module	EMS 8911
or prime mover/dynamometer	EMS 8960
Hand-held tachometer	EMS 8920
Single-phase wattmeter module	EMS 8431
Capacitor run motor module	EMS 8253
Ohmmeter (supplied by student)	

Discussion

The capacitor start capacitor run motor, shown in Figure 44-1, has increased in popularity over the years. This type of split-phase motor does not disconnect the start windings from the circuit when it is running. This eliminates the need for a centrifugal switch or starting relay to disconnect the start windings from the circuit when the motor reaches about 75% of its full speed. This motor does not have as good a starting torque as the capacitor start induction run motor, however. The capacitor is designed to be used by the motor during running and is, therefore, smaller than would be required for starting. For this reason, the capacitor start capacitor run motor will sometimes use an extra capacitor to aid in starting. When this is done, the start capacitor is connected in parallel with the run capacitor. During the time of starting, both of these capacitors are connected in the circuit, as shown in Figure 44-2. When the motor has accelerated to about 75% of full speed, the start capacitor is disconnected from the circuit. If the motor is an open type, the start capacitor will be disconnected by a centrifugal switch. If the motor is hermetically sealed, a starting relay will be used to disconnect the start capacitor.

Since the run capacitor remains in the circuit at all times, an AC oil-filled capacitor is generally used on motors of about 1/4 hp and larger instead of an AC electrolytic capacitor. Small capacitor start capacitor run motors used to operate ceiling fans and small blowers generally use an AC electrolytic capacitor, however, because of space limitations. The run capacitor does perform power factor correction for the motor. This type of motor is very smooth in operation and produces little hum or noise.

FIGURE 44-1 Capacitor start capacitor run motor. (Courtesy of Lab-Volt® Systems, Inc.)

The windings of this motor are different from those of the capacitor start induction run motor also. In the capacitor start induction run motor, the start winding is made of much smaller wire than the run winding. The start and run windings of the capacitor start capacitor run motor are closer to being the same size. The run capacitor is used to produce the phase angle shift between the current in the run winding and the current in the start winding. Because the run capacitor and start winding remain in the circuit at all times, this motor operates as a two-phase motor, which accounts for its smooth operation and low noise. The direction of rotation of this motor can be changed by reversing either the run winding leads or the start winding leads.

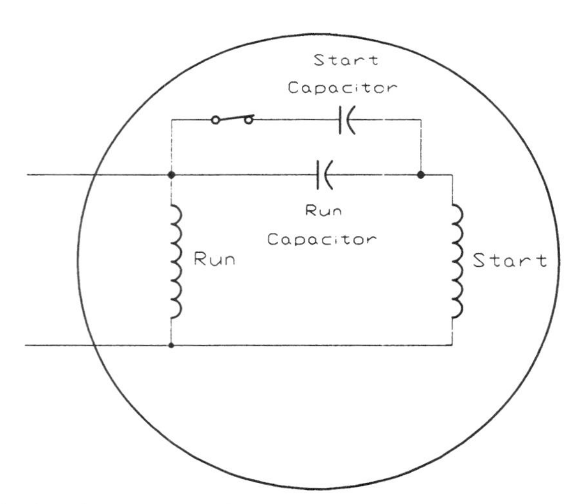

FIGURE 44-2 Capacitor start capacitor run motor with additional starting capacitor

Name ______________________________ Date ______________

Procedure

1. Remove the capacitor run motor module from the mobile console.
2. Examine the module and locate the following items:
   A. The run capacitor. Notice that this capacitor is an AC oil-filled capacitor and not an AC electrolytic capacitor like that used with the capacitor start induction run motor.
   B. The run and start windings. Observe that these windings are closer to being the same size than those of the capacitor start induction run motor. Also, this motor does not contain a centrifugal switch to disconnect the start winding.
3. Replace the motor in the mobile console.
4. Using an ohmmeter, measure the resistance of the run winding across terminals 1 and 2.

   ____________ Ω
5. Measure the resistance of the start winding across terminals 3 and 4.

   ____________ Ω
6. Connect the circuit shown in Figure 44-3 if using the electrodynamometer. If the prime mover/dynamometer is being used, connect the 24 VAC low-voltage connection to the power supply.
7. Connect the motor to the dynamometer with the timing belt and set the dynamometer control knob to the full counterclockwise position.

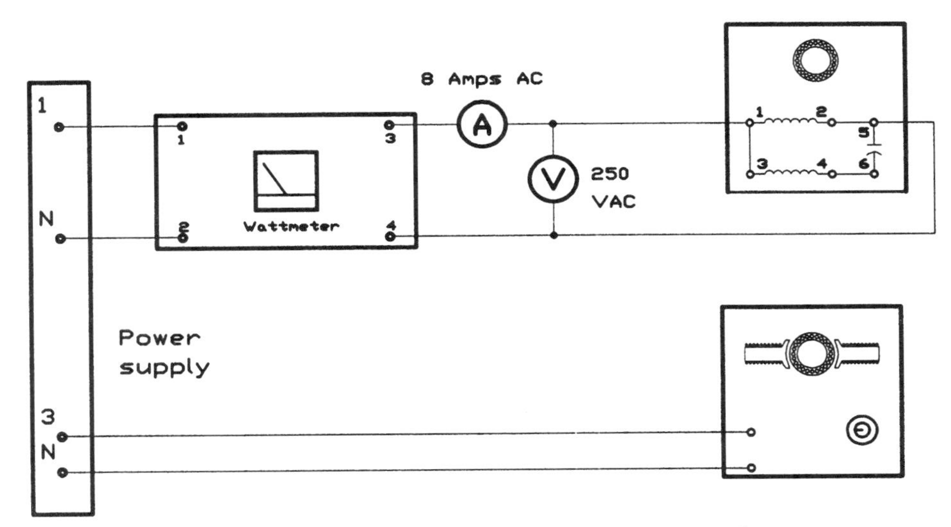

FIGURE 44-3 Capacitor start capacitor run motor connection

8. Turn on the power supply and fill in the chart shown in Figure 44-4 using the following formulas.

   The horsepower can be computed using the formula

$$hp = \frac{6.28 \times RPM \times L \times P}{33,000}$$

   where 6.28 is a constant
   RPM = speed in revolutions per minute
   L = length in feet (0.08333)
   P = pounds of force
   33,000 is a constant

   The output power can be computed by using the formula

$$\text{output power} = \text{output horsepower} \times 746$$

   The efficiency can be computed by using the formula

$$\text{eff.} = \frac{\text{output power}}{\text{input power}} \times 100$$

   The apparent power (VA) can be computed using the formula

$$VA = E \times I$$

   The power factor can be computed using the formula

$$PF = \frac{W(\text{input})}{VA} \times 100$$

9. **Turn off the power supply and disconnect the circuit.**
10. Return the components to their proper place.

Load torque (lb–in.)	Armature current (amps)	Speed (RPM)	Input power (watts)	Output horsepower (hp)	Output power (watts)	Eff. (%)	Apparent power (VA)	PF (%)
0								
2								
4								
6								
8								
10								
12								

FIGURE 44-4

Review Questions

1. Why does the capacitor start capacitor run motor not contain a centrifugal switch?

 ____

2. Does this type of motor provide power factor correction?

 ____

3. Why do some capacitor start capacitor run motors contain a starting capacitor as well as a running capacitor?

 ____

4. What type of device is used to disconnect the starting capacitor from the circuit in a hermetically sealed motor?

 ____

5. How is the direction of rotation reversed?

 ____

Exercise 45

The Universal Motor: Part 1

Objectives

After completing this lab you should be able to:

- Discuss the operation of a universal motor.
- Connect a universal motor for operation on AC or DC current.
- Make measurements with test instruments.

Materials and Equipment

Power supply module	EMS 8821
Universal motor module	EMS 8254
Hand-held tachometer	EMS 8920
AC voltmeter module	EMS 8426
DC metering module	EMS 8412
Electrodynamometer	EMS 8911
or prime mover/dynamometer	EMS 8960

Discussion

The universal motor is often referred to as an AC series motor. This motor is very similar to a DC series motor in its construction. The universal motor, however, has the addition of a compensating winding. If a DC series motor were connected to alternating current, the motor would operate very poorly. There are two main reasons for this. First, the armature windings would have a large amount of inductive reactance if they were connected to alternating current. Second, the field poles of most DC machines contain solid metal pole pieces. If the field were connected to AC, a large amount of power would be lost to eddy current induction in the pole pieces. Universal motors contain a laminated core to help prevent the problem of eddy current induction. The compensating winding is wound around the stator and functions to counteract the inductive reactance in the armature winding.

The universal motor is so named because it can be operated on AC or DC voltage. When the motor is operated on direct current, the compensating winding is connected in series with the series field winding.

CONNECTING THE COMPENSATING WINDING FOR ALTERNATING CURRENT

When the universal motor is operated with AC power, the compensating winding can be connected in two ways. If it is connected in series with the armature, as shown in Figure 45-1, it is known as conductive compensation.

The compensating winding can also be connected by shorting its leads together, as shown in Figure 45-2. When connected in this manner, the winding acts like a shorted secondary winding of a transformer. Induced current permits the winding to operate when connected in this manner. This connection is known as inductive compensation. Inductive compensation cannot be used when the motor is connected to DC.

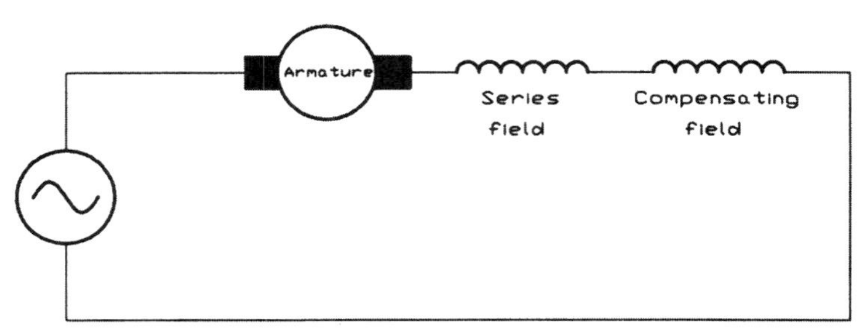

FIGURE 45-1 Conductive compensation

THE NEUTRAL PLANE

Because the universal motor contains a wound armature, commutator, and brushes, the brushes should be set at the neutral plane position. This can be done in the universal motor in a manner similar to that of setting the neutral plane of a DC machine. When the brushes are set to the neutral plane position in a universal motor, either the series field or compensating winding can be used. The brushes are set to the neutral plane position using the series field winding, by connecting alternating current to the armature leads, as shown in Figure 45-3. A voltmeter is connected to the series winding. Voltage is then applied to the armature. The brush position is then moved until the voltmeter connected to the series field reaches a null position. (The null position is reached when the voltmeter reaches its lowest point.)

If the compensating winding is used to set the neutral plane, alternating current is again connected to the armature and a voltmeter is connected to the compensating winding, as shown in Figure 45-4. Alternating current is then applied to the armature. The brushes are then moved until the voltmeter indicates its highest, or peak, voltage.

SPEED REGULATION

The speed regulation of the universal motor is very poor. Because this motor is a series motor, it has the same poor speed regulation as a DC series motor. If the universal motor is connected to a light load or no load, its speed is almost unlimited. It is not unusual for this motor to be operated at several thousand RPM. Universal motors are used in a number of portable appliances where high horsepower and light weight are needed, such as drill motors, skill saws, and vacuum cleaners. The universal motor is able to produce a high horsepower for its size and weight because of its high operating speed. Universal motors can generally be identified by their commutator and brushes.

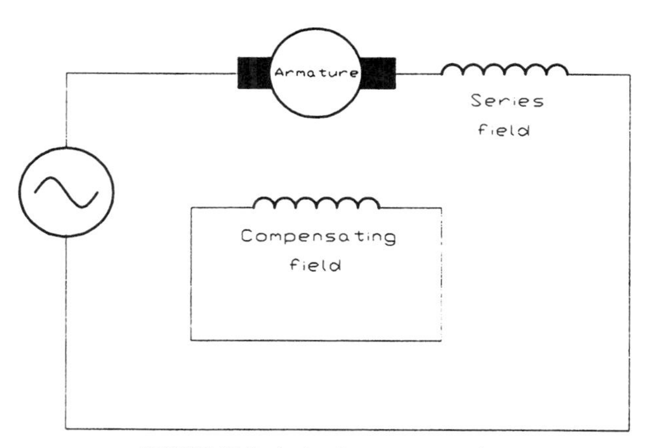

FIGURE 45-2 Inductive compensation

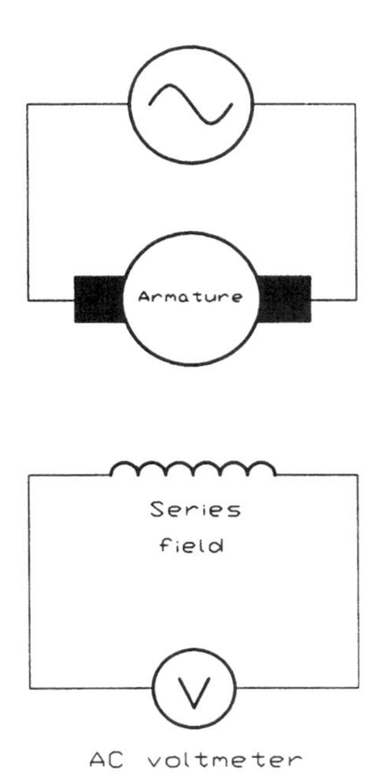

FIGURE 45-3 Setting the neutral plane using the series field

Only two types of AC motors have a commutator and brushes, the universal motor and the repulsion motor.

DIRECTION OF ROTATION

The direction of rotation of the universal motor can be changed in the same manner as changing the direction of rotation of a DC series motor. To change the direction of rotation, change the armature leads with respect to the field leads.

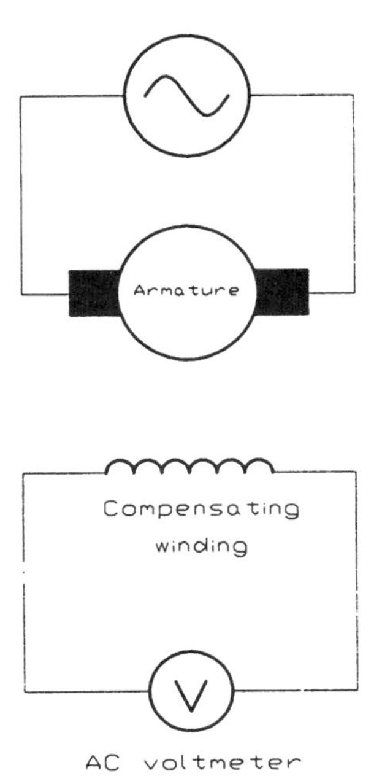

FIGURE 45-4 Setting the neutral plane using the compensating winding

Name ______________________________ Date ______________

Procedure

1. Remove the universal motor module shown in Figure 45-5 from the mobile console.
2. Locate the following items:

   A. The armature

   B. The commutator

   C. The brushes

   D. The brush position set control

   E. The stator windings
3. Replace the universal motor module in the mobile console.
4. Connect the motor to the dynamometer with the timing belt. Set the dynamometer control knob to the full counter-clockwise position.
5. Connect the circuit shown in Figure 45-6.
6. Turn on the power supply and apply a voltage of **50 V** to the armature of the universal motor.
7. Open the front panel cover of the universal motor module, and move the brush position control from one side to the other. Notice the change of voltage induced in the series field winding. Set the brush position control to a point where the voltmeter connected to the series field is null or at its lowest value. This is the neutral plane position of the motor.
8. **Turn off the power supply.**
9. Close the front cover of the motor module and connect the circuit shown in Figure 45-7.

FIGURE 45-5 Universal motor module. (Courtesy of Lab-Volt® Systems Inc.)

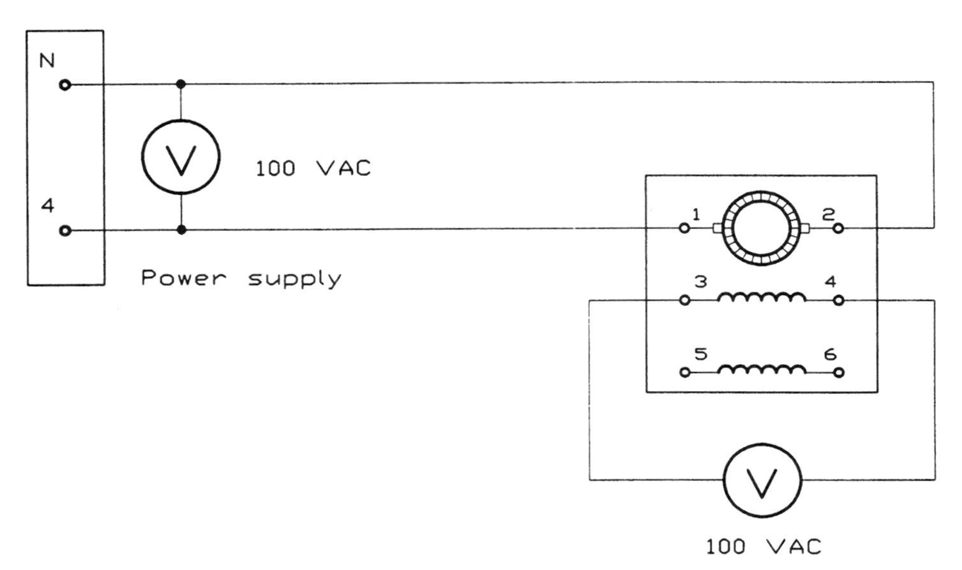

FIGURE 45-6 Setting the neutral plane using the series field winding

10. Turn on the power supply and apply a voltage of **50 V** to the armature.
11. Again move the brush position control from side to side and observe the voltage reading on the voltmeter connected to the compensating field. Set the brush position control to a point where the voltmeter is at its peak value. This is the neutral plane of the motor. Notice that this is the same position that was found when using the series field.
12. **Turn off the power supply** and close the front cover of the motor module.
13. Connect the circuit shown in Figure 45-8 if using the electrodynamometer. If the prime mover/dynamometer is being used, connect the 24 VAC low-voltage connection to the power supply. In this exercise the universal motor will operate as a DC machine. Notice that the compensating winding has not been connected.

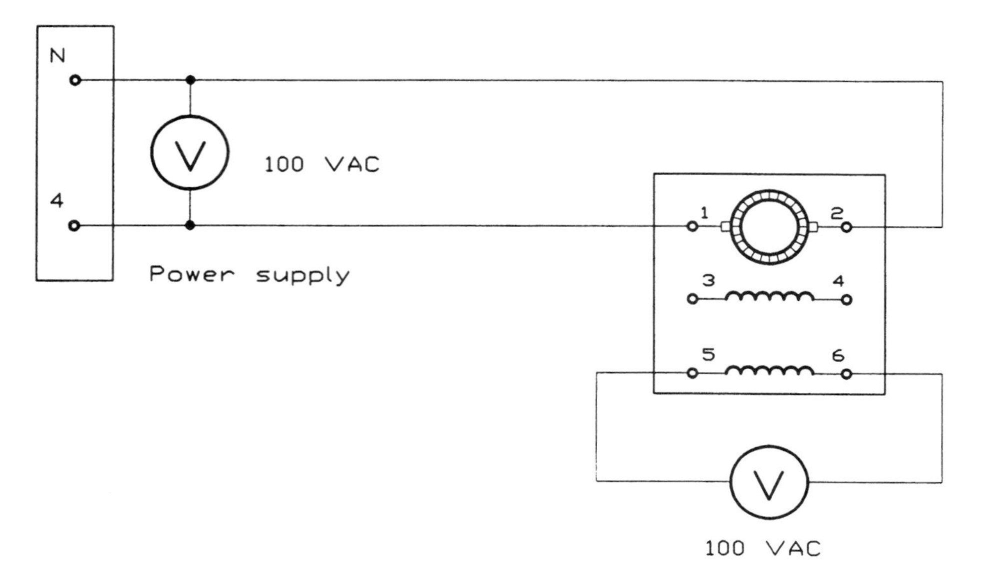

FIGURE 45-7 Setting the neutral plane using the compensating winding

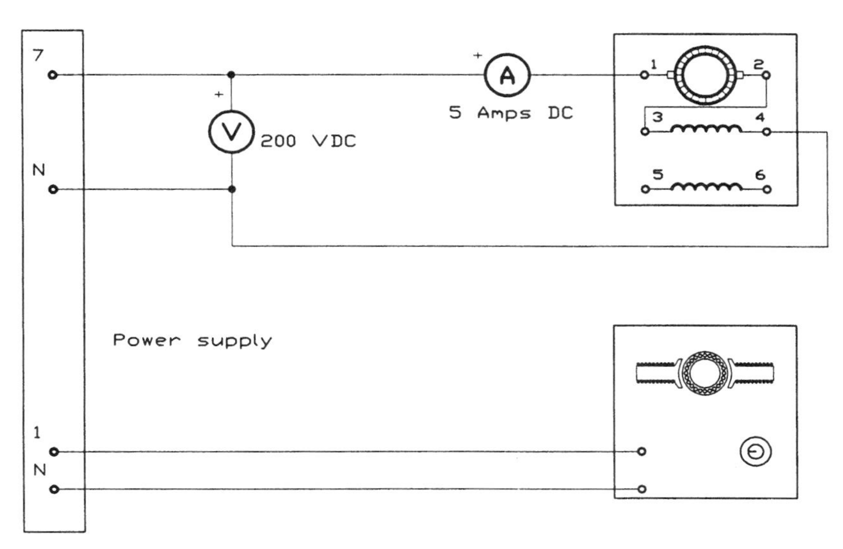

FIGURE 45-8 Operating a universal motor on DC current without compensation

14. Turn on the power supply and adjust the output voltage for a value of 120 VDC. This voltage should be maintained throughout the experiment.

15. Fill in the chart shown in Figure 45-9 by using test instruments and the following formulas. (Note: It may not be possible to reach the full load of 8 lb.-in. because of the unstable nature of a universal motor operated without a compensating winding.)

$$hp = \frac{6.28 \times RPM \times L \times P}{33,000}$$

where 6.28 is a constant
RPM = speed in revolutions per minute
L = length in feet (0.08333)
P = pounds of force
33,000 is a constant

The output power can be computed by using the formula

$$\text{output power} = \text{output horsepower} \times 746$$

Load torque (lb-in.)	Armature current (amps)	Speed (RPM)	Input power (watts)	Output horsepower (hp)	Output power (watts)	Eff. %
0						
2						
4						
6						
8						

FIGURE 45-9

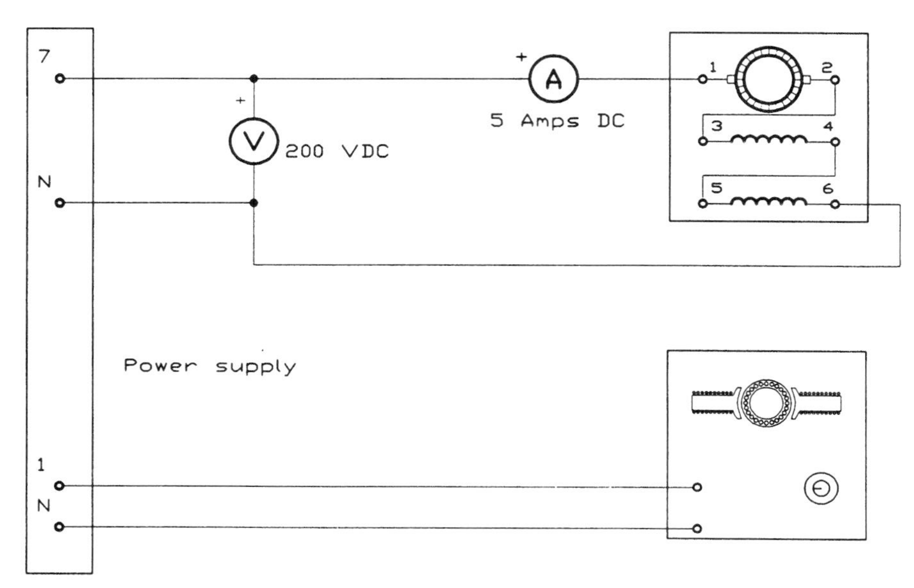

FIGURE 45-10 Operating a universal motor on DC current with conductive compensation

Because this is a DC circuit, the input power can be computed by multiplying the line current and applied voltage.

$$\text{input power} = E \times I$$

The efficiency can be computed by using the formula

$$\text{eff.} = \frac{\text{output power}}{\text{input power}} \times 100$$

16. **Return the voltage to 0 V and turn off the power supply.**
17. Connect the circuit shown in Figure 45-10. In this circuit, the compensating winding has been connected in series with the series winding.
18. Fill in the chart shown in Figure 45-11.
19. **Return the voltage to 0 V and turn off the power supply.**
20. Disconnect the circuit and return the components to their proper place.

Load torque (lb-in.)	Armature current (amps)	Speed (RPM)	Input power (watts)	Output horsepower (hp)	Output power (watts)	Eff. %
0						
2						
4						
6						
8						

FIGURE 45-11

Review Questions

1. Why is the AC series motor often referred to as a universal motor?

 ___

2. What is the function of the compensating winding?

 ___

3. How is the direction of rotation of the universal motor reversed?

 ___

4. When the motor is connected to DC voltage, how must the compensating winding be connected?

 ___

5. Explain how to set the neutral plane position of the brushes using the series field.

 ___

6. Explain how to set the neutral plane position using the compensating winding.

 ___

Exercise 46

The Universal Motor: Part 2

Objectives

After completing this lab you should be able to:

- Discuss the operation of a universal motor connected to an AC circuit.
- Evaluate the performance of a universal motor when the compensating winding is used and when it is not used.
- Connect a universal motor and make measurements using test equipment.

Materials and Equipment

Power supply module	EMS 8821
Universal motor module	EMS 8254
Hand-held tachometer	EMS 8920
Single-phase wattmeter module	EMS 8431
AC ammeter module	EMS 8425
AC voltmeter module	EMS 8426
Electrodynamometer module	EMS 8911
or prime mover/dynamometer	EMS 8960

Discussion

In this laboratory experiment, the universal motor will be connected for operation on AC power. The motor will first be operated without the compensating winding connected in the circuit. In the next section, the compensating winding will be connected to provide conductive compensation, and in the final section the compensating winding will be connected to provide inductive compensation.

Name ______________________________ Date ______________

Procedure

1. Connect the circuit shown in Figure 46-1 if using the electrodynamometer. If the prime mover/dynamometer is being used, connect the 24 VAC low-voltage connection to the power supply.
2. Connect the motor to the dynamometer with the timing belt.
3. Turn on the power supply and adjust the output voltage for a value of 120 VAC. This voltage should be maintained throughout the experiment.
4. Fill in the chart shown in Figure 46-2 by using measuring instruments and the formulas shown.

$$\text{hp} = \frac{6.28 \times \text{RPM} \times L \times P}{33,000}$$

where 6.28 is a constant
RPM = speed in revolutions per minute
L = length in feet (0.08333)
P = pounds of force
33,000 is a constant

The output power can be computed by using the formula

$$\text{output power} = \text{output horsepower} \times 746$$

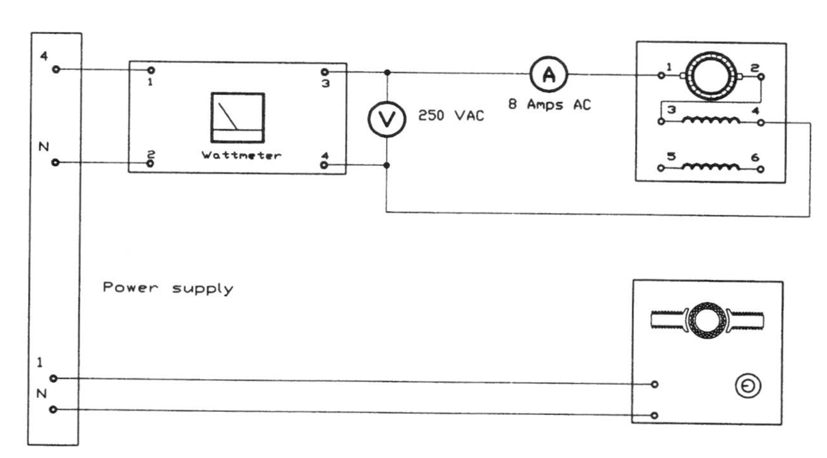

FIGURE 46-1 Universal motor operated on AC current without compensation

Load torque (lb-in.)	Armature current (amps)	Speed (RPM)	Input power (watts)	Output horsepower (hp)	Output power (watts)	Eff. %
0						
2						
4						

FIGURE 46-2

The efficiency can be computed by using the formula

$$\text{eff.} = \frac{\text{output power}}{\text{input power}} \times 100$$

5. **Return the voltage to 0 V and turn off the power supply.**
6. Connect the circuit shown in Figure 46-3. In this circuit, the universal motor has been reconnected in such a manner that it is conductively compensated.
7. Turn on the power supply and adjust the output voltage for a value of 120 VAC. Maintain this voltage value throughout the experiment.
8. Fill in the values of the chart shown in Figure 46-4.
9. **Return the voltage to 0 V and turn off the power supply.**
10. Connect the circuit shown in Figure 46-5. In this circuit the motor is connected for inductive compensation.
11. Turn on the power supply and adjust the output voltage for a value of 120 V. Maintain this voltage throughout the experiment.
12. Fill in the chart in Figure 46-6.
13. **Return the voltage to zero and turn off the power supply.**
14. Disconnect the circuit and return the components to their proper place.

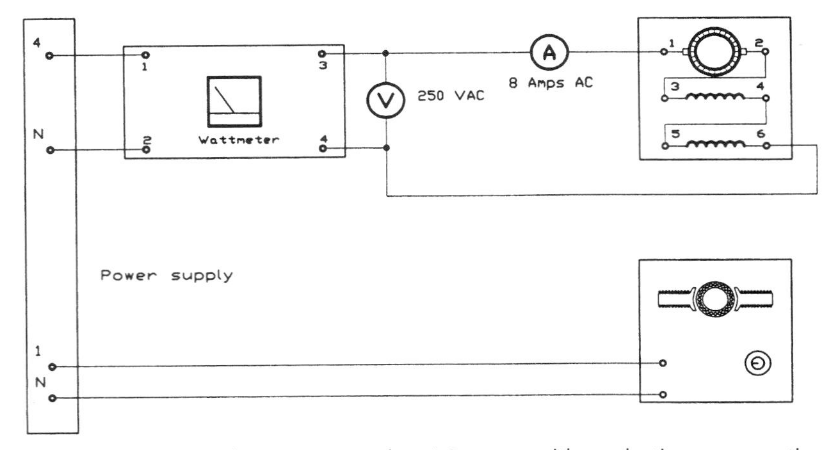

FIGURE 46-3 Universal motor operated on AC current with conductive compensation

Load torque (lb-in.)	Armature current (amps)	Speed (RPM)	Input power (watts)	Output horsepower (hp)	Output power (watts)	Eff. %
0						
2						
4						
6						
8						

FIGURE 46-4

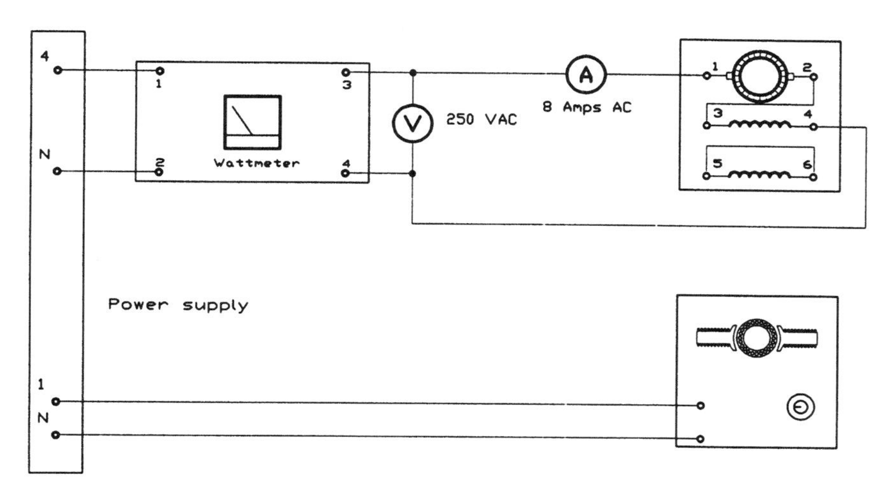

FIGURE 46-5 Universal motor operated on AC current with inductive compensation

Load torque (lb-in.)	Armature current (amps)	Speed (RPM)	Input power (watts)	Output horsepower (hp)	Output power (watts)	Eff. %
0						
2						
4						
6						
8						

FIGURE 46-6

Review Questions

1. Explain the difference between conductive and inductive compensation.

   Conductive ______________________________

   Inductive ______________________________

2. Did the motor operate more efficiently when connected for conductive or inductive compensation?

   ____

Ohmic values obtained from resistive, capacitive, and inductive load modules. (All sections connected in parallel)			
Total Value (Ohms)	First Section Closed Switches (Ohms)	Second Section Closed Switches (Ohms)	Third Section Closed Switches (Ohms)
1200	1200	None	None
600	600	None	None
400	1200 & 600	None	None
300	300	None	None
240	1200 & 300	None	None
200	600 & 300	None	None
171.4	1200 & 600 & 300	None	None
150	1200 & 600 & 300	1200	None
133,3	1200 & 600 & 300	600	None
120	600 & 300	300	None
109.1	1200 & 600 & 300	300	None
100	300	300	300
92.3	1200 & 600 & 300	600 & 300	None
85.7	1200 & 600 & 300	1200 & 600 & 300	None
80	1200 & 600 & 300	1200 & 600 & 300	1200
75	1200 & 600 & 300	1200 & 600 & 300	600
70.6	1200 & 600 & 300	600 & 300	300
66.7	1200 & 600 & 300	1200 & 600 & 300	300
63.1	1200 & 600 & 300	1200 & 600 & 300	1200 & 300
60	1200 & 600 & 300	1200 & 600 & 300	600 & 300
57.1	1200 & 600 & 300	1200 & 600 & 300	1200 & 600 & 300

Chart of Ohmic Values